Vladimir Kirtskhalia

Applied problems of gas and hydrodynamics

Vladimir Kirtskhalia

Applied problems of gas and hydrodynamics

LAP LAMBERT Academic Publishing

Impressum / Imprint

Bibliografische Information der Deutschen Nationalbibliothek: Die Deutsche Nationalbibliothek verzeichnet diese Publikation in der Deutschen Nationalbibliografie; detaillierte bibliografische Daten sind im Internet über http://dnb.d-nb.de abrufbar.

Alle in diesem Buch genannten Marken und Produktnamen unterliegen warenzeichen-, marken- oder patentrechtlichem Schutz bzw. sind Warenzeichen oder eingetragene Warenzeichen der jeweiligen Inhaber. Die Wiedergabe von Marken, Produktnamen, Gebrauchsnamen, Handelsnamen, Warenbezeichnungen u.s.w. in diesem Werk berechtigt auch ohne besondere Kennzeichnung nicht zu der Annahme, dass solche Namen im Sinne der Warenzeichen- und Markenschutzgesetzgebung als frei zu betrachten wären und daher von jedermann benutzt werden dürften.

Bibliographic information published by the Deutsche Nationalbibliothek: The Deutsche Nationalbibliothek lists this publication in the Deutsche Nationalbibliografie; detailed bibliographic data are available in the Internet at http://dnb.d-nb.de.

Any brand names and product names mentioned in this book are subject to trademark, brand or patent protection and are trademarks or registered trademarks of their respective holders. The use of brand names, product names, common names, trade names, product descriptions etc. even without a particular marking in this work is in no way to be construed to mean that such names may be regarded as unrestricted in respect of trademark and brand protection legislation and could thus be used by anyone.

Coverbild / Cover image: www.ingimage.com

Verlag / Publisher:
LAP LAMBERT Academic Publishing
ist ein Imprint der / is a trademark of
OmniScriptum GmbH & Co. KG
Heinrich-Böcking-Str. 6-8, 66121 Saarbrücken, Deutschland / Germany
Email: info@lap-publishing.com

Herstellung: siehe letzte Seite /
Printed at: see last page
ISBN: 978-3-659-74802-8

Copyright © 2015 OmniScriptum GmbH & Co. KG
Alle Rechte vorbehalten. / All rights reserved. Saarbrücken 2015

Vladimir G. Kirtskhalia
I. Vekua Sukhumi Institute of Physics and Technology (SIPT)

Applied problems of gas and hydrodynamics

Tbilisi

2015

Table of Contents

Introduction..3

CHAPTER I
Thermodynamic characteristic of the Earth's atmosphere

The problems 4
§ 1. Modern Representations about Speed of
  Sound in Atmosphere of the Earth 5
§ 2. Influences of heterogeneity of medium on the generation
  and propagation of sound 8
§ 3. Speed of Sound in Atmosphere of the Earth 10
§ 4. The coefficient of thermal expansion in the Earth's atmosphere 13
  Results...18

CHAPTER II
Compressibility or incompressibility as properties of an inhomogeneous medium

The Problems... 18
§ 1. Internal gravity waves, as an example of the
  incorrectness of modern hydrodynamics...19
§ 2. Equation of mass continuity for inhomogeneous medium..............20
§ 3. Changing of densities of the flows of some of the physical
  quantities in inhomogeneous medium... 23
  Results...25

CHAPTER III
Peculiarities of sound waves in inhomogeneous medium

The Problems... 26
§ 1. Generalized equation of gravitational waves
  in moving inhomogeneous medium..26
§ 2. Sound Wave as a Particular Case of the
  Gravitational Waves..29
§ 3. Peculiarities of Sound Wave in Moving Medium............................33
  Results...36

CHAPTER IV
Disadvantages of theory of tangential discontinuity and how to overcome them

The Problems...37
§ 1. Incorrectness of the modern theory of hydrodynamic
  tangential discontinuity...37
§ 2. The correct solution to the task of Landau.......................................43
§ 3. Incorrectness of modern linear theory
  of surface capillary-gravity waves..47
§ 4. The correct solution of the task of the capillary waves...................52
  Results...56
  Conclusion...57
  References 60

Introduction

The book is intended for specialist in the field of gas and hydrodynamics. Creation of the book was motivated by longstanding scientific researching of applied problems carried out by the author in Sukhumi Institute of Phisics and Technology in this important sub-discipline of physics. Analysis of monographs of well-known authors like Landau L.D. and Lifchitz E.N., Pierse A.D. , Gossard E. E. and Hooke W. H., Stoker J.J. Kowalik Z, etc. being of the handbooks for scientists of this field, has shown that most of applied tasks in gas and hydrodynamics are being solved incorrectly. This book is a compilation of published in the last three years of studies in which identified existing incorrectnees at solving applied problems of gas and hydrodynamics and the ways to overcome them are given. We hope that the experts in this field of science will find a lot of useful tips and recommendations for their research.

Chapter I deals with the fitness of existing theory of gas and hydrodynamic with respect to inhomogeneous medium. It is shown that many characteristic thermodynamic parameters and criteria, applied to any medium, are only true for homogeneous medium and for non-homogeneous medium they should be improved This is particularly true to the definition of such widespread phenomenon in nature as sound and to determination of its speed. According to our results, in inhomogeneous medium sound has two mechanisms of generation and propagation: adiabatic and isobaric, unlike homogeneous medium where there is only the adiabatic mechanism. Each of these mechanism of perturbation of dencity has its own propagation speed which we call adiabatic and isobaric speeds of sound. Real value of squared of sound speed is the reduced from squares of these two speeds which leads to dependence of sound speed on altitude in the atmosphere of the Earth.

In chapter II, we disprove the existing view that such important criteria characterizing of medium as compressibility and incompressibility are not properties of medium, but they arise from conditions of a particular task. It is shown that such understanding of criteria leads to paradoxical results. It is incontrovertibly stated that compressibility or incompressibility are directly connected with homogeneity or inhomogeneity of medium , which are the properties of medium and depend on mechanisms of generation and propagation of sound. If in the medium, adiabatic mechanism of generation of sound prevails, then the medium is homogeneous and consequently, it is compressible, but if an isobaric mechanism prevails, the medium is non-homogeneous and consequently, it is incompressible, Thus, terms "compressible medium" and "'incompressible medium", which had mechanical meaning, acquire thermodynamical meaning. For example, it is shown that water and iron, in thermodynamic sense, are much more compressible than air in the upper layers of atmosphere.

In chapter III, on the basis of generalized equation of continuity of mass and new determination of criteria of compressibility and incompressibility, generalized equation of linear gravity wave is obtained, from which can be received an equation for any special case. It is shown that from this equation, we get the equation of sound wave in the atmosphere of Earth and consequently, the common opinion that the sound wave is not gravitational, is disproved. Effects of gravitational field on sound wave are revealed. It is shown that Doppler's Law and the principle of relativity of motion, that were defined for homogeneous medium, fair to the sound wave in highly heterogeneous stationary medium. In moving heterogeneous medium, sound propagation is impossible.

In chapter IV considered problematic issues of theory of linear surface gravitational waves. It is shown that the theory of hydrodynamic tangential discontinuity whereby which many problems of surface gravity waves are solved, has many drawbacks. Namely, this theory is based on assumptions of potentiality of motion and incompressibility of liquids, which are not applicable in the gravitational field of the earth. It is proved that linear theory of tangential shear describes only capillary waves, on which gravitational field does not influence. Thus, the capillary- gravitational wave, known since the time of Kelvin, does not exist. Does not exist also the condition, limiting the length of capillary wave, and consequently, the assertion that the capillary waves propagate only in deep water is wrongful.

CHAPTER I

Thermodynamic characteristic of the Earth's atmosphere

The problems

Sound speed is a characteristic quantity of the medium which is included in the system of hydrodynamic (gas-dynamic) equations and plays a significant role in study of wave processes within it. Therefore, correct determination of its value is crucial in adequate description of generation and propagation of waves in media. According to the modern theory sound wave is considered as propagation of density perturbation at variation of mass in constant volume without heat transfer i.e. adiabatically and therefore the speed of sound propagation is called adiabatic speed of sound. According to this representation the speed of sound in atmosphere is calculated according to the formula $C^2 = \gamma RT/M$, i.e. depends only on temperature and does not depend on altitude. This result is doubtful, since it implies that under the identical temperature conditions at the sea level and at an altitude of 100 km, where air density is 10^7 times less, the speed of sound should have the same meaning. Figures 1 and 2 demonstrate graphs of dependence of temperature and sound speed in atmosphere of the Earth on altitude in the range from of 0 to 85 km [1] which almost coincide. The same paper presents the table demonstrating numerical meanings of sound speed of the same interval and it is indicated that these meanings are calculated by means of the mentioned formula (Figure 13 shows the variation with altitude of the computed speed of sound). Internet provides the same meanings of sound speed [2]. This result of existing theory of acoustic waves is contrasting with the fundamental principle of physics. According to this principle all characteristic parameters of inhomogeneous medium must also be inhomogeneous, i.e. must depend on space coordinates. Since the gravity field of earth influences on the atmosphere it is essentially inhomogeneous medium, speed of sound in which must clearly depend on z coordinate (altitude) $C = C(z,T)$. Although this contradiction is evident to everyone, it still exists.

The coefficient of thermal expansion β is the same characteristics of the air, as well as the speed of sound and consequently, according to the above mentioned principle, in the Earth's atmosphere it is must depend as on the initial temperature as well as the height, i.e. $\beta = \beta(z,T)$.

Defining the analytical type of functions $C = C(z,T)$ and $\beta = \beta(z,T)$ is an extremely important goal of applied physics.

§1. . Modern Representations about Speed of Sound in Atmosphere of the Earth

In this paragraph we deal with the existing theory of sound waves and with the most typical examples which demonstrate that the determination of sound speed in the Earth's atmosphere is not complete.

Motion of frictionless liquid (or gas) in the gravitational field of the Earth is described by the equation of Euler:

$$\rho\left[\frac{\partial \vec{v}}{\partial t} + (\vec{V}\nabla)\vec{V}\right] = -\nabla P + \rho \vec{g} \quad (1.1.1)$$

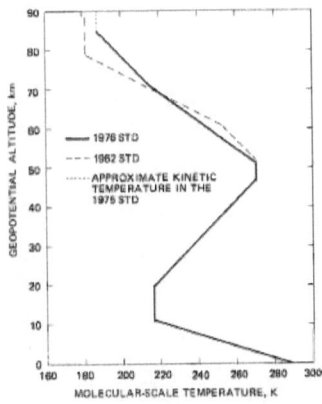
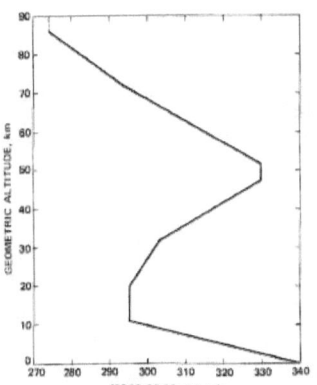

Figure 1. Molecular-scale temperature as a function of geopotencial altitude.

Figure 2. Speed of sound as a function of geometrical altitude.

which is solved in conjunction with the equations of continuity of mass

$$\frac{\partial \rho}{\partial t} = -div(\rho \vec{V}) \quad (1.1.2)$$

and adiabatic equation

$$\frac{\partial s}{\partial t} + (\vec{V}\nabla s) = 0 \quad (1.1.3)$$

Exact solution of this system in general case is impossible, so scientists uses approximate method of small disturbances, when variables who are included in the system of equations (1.1.1-3) are presented as the sum of their stationary and disturbed values $f(\vec{r},t) = f_0(\vec{r}) + f'(\vec{r},t)$, where $f'/f_0 < 1$ and members of the order $(f')^n$, where $n > 1$ are neglected. Velocity of the fluid (gas) in the acoustic wave is considered as small value. For the pressure and density we will have $p = P_0 + p'$, $\rho = \rho_0 + \rho'$ and therefore after linearization of the equation (1.1.1) we obtain

$$\rho_0 \frac{\partial \vec{V}}{\partial t} = -\nabla(P_0 + p') + (\rho_0 + \rho')\vec{g} \qquad (1.1.4)$$

Let's consider two examples of application of equation (1.1.4) for sound waves which give clear representation of inadequacy of the current theory of sound waves in the Earth atmosphere.

In monographs [3,4] the Equation (1.1.4) is given as follows:

$$\rho_0 \frac{\partial \vec{V}}{\partial t} = -\nabla p' \qquad (1.1.5)$$

Here it is implied that $g = 0$ and $\nabla P_0 = 0$ i.e. gravitational field does not affect the medium. The Equation (1.1.5) is solved together with the linearized equation of the continuity of mass

$$\frac{\partial \rho'}{\partial t} + \rho_0 div \vec{V} = 0 \qquad (1.1.6)$$

when relation between perturbations of pressure and density are given by the formula

$$p' = \left(\frac{\partial P_0}{\partial \rho_0}\right)_s \rho' \qquad (1.1.7)$$

The index s means that the derivative undertakes at constant entropy. The potential of speed is defined as $\vec{V} = \nabla \varphi$ after which the equations (1.1.5-7) lead to the wave equation:

$$\frac{\partial^2 \varphi}{\partial t^2} - C_s^2 \Delta \varphi = 0 \qquad (1.1.8)$$

where quantity

$$C_s = \sqrt{\left(\frac{\partial P_0}{\partial \rho_0}\right)_s} \qquad (1.1.9)$$

defined as adiabatic speed of sound. Considering that the air is ideal gas ($p = nkT$), and dependence between pressure and density in adiabatic process is defined by the following relation

$$\frac{P_0}{P_0^{(1)}} = \left(\frac{\rho_0}{\rho_0^{(1)}}\right)^{\gamma} \quad (1.1.10)$$

where $P_0^{(1)}$ and $\rho_0^{(1)}$ are initial values of pressure and density, and adiabatic index $\gamma = c_p/c_v = 1,4$ is the ratio of specific heats at constant pressure and volume respectively, from (1.2.9), for speed of sound following expression is derived:

$$C_s = \sqrt{\gamma \frac{P_0}{\rho_0}} = \sqrt{\gamma \frac{k_B T}{m_0}} = \sqrt{\gamma \frac{RT}{M}} \quad (1.1.11)$$

Here $k_B = 1,38 \times 10^{-23} \, J/{}^\circ K$ is Boltzmann's constant, $R = 8,314 \, J/(mol\,{}^\circ K)$ is a gas constant, $m_0 = 4,81 \times 10^{-26} \, kg$ is mass of one air molecule and $M = 29 \times 10^{-3} \, kg/mol$ -mass of one mol air.

In monographies [5-7], the condition of static balance

$$\nabla P_0 = \rho_0 \vec{g} \quad (1.1.12)$$

is applied to the equation (1.1.4) and then it becomes:

$$\rho_0 \frac{\partial \vec{V}}{\partial t} = -\nabla p' + \rho' \vec{g} \quad (1.1.13)$$

The authors think that gravitational acceleration g does not influence high-frequency sound fluctuations and the equation (1.1.13) passes into the equation (1.1.5). They supplement the equations (1.1.5) and (1.1.6) with the equation of adiabatic process for ideal gas

$$\frac{\partial p'}{\partial t} - \gamma \frac{RT}{M} \frac{\partial \rho'}{\partial t} = 0 \quad (1.1.14)$$

and representing all perturbed quantities in the form of a plane wave $\exp[i(kx - \omega t)]$ and get from these three equations the wave equation relative to x component of perturbed speed

$$\frac{\partial^2 u}{\partial t^2} - \gamma \frac{RT}{M} \frac{\partial^2 u}{\partial x^2} = 0 \quad (1.1.15)$$

from which they obtain expression (1.1.11) for the speed of sound. This expression presently is applied for both bottom ($0 \leq z \leq 11$ km) and top ($11 < z \leq 86$ km) atmospheres.

Thus, in both cases authors try to define speed of sound in atmosphere of the Earth from the wave equation for a plane wave in homogeneous medium, which depends only on temperature. It is well known, that to this equation satisfy the tensions of electric and magnetic fields of electromagnetic wave in vacuum for which the vacuum is a homogeneous medium.

Consequently the velocity of light in it is homogeneous. Earth's atmosphere is essentially heterogeneous medium and all its thermodynamic parameters according to the fundamental principle of physics must depend on coordinates. Since the speed of sound is characteristic parameter of the medium, it must depend on coordinates too. Proceeding from the aforementioned, formulation and solution of the problem in the monographs [3,4] look correct, however they have no relation to a sound wave in atmosphere of the Earth.

As for the results obtained in monographs [5-7], it can be easily shown that it is incorrect. Actually, in Euler's linearized equation the authors apply condition of statistic balance (1.1.12) according to which

$$P_0(z) = P_0^0 - \rho_0^0 gz \tag{1.1.16}$$

Here P_0^0 and ρ_0^0 is values of pressure and density on a sea level. Expression (1.1.16) turns out from the general formula of distribution of pressure in atmosphere of the Earth

$$P_0 = n_0 k_B T = \frac{\rho_0 k_B T}{m_0} = \frac{\rho_0^{(0)}}{m_0} k_B T \exp\left(-\frac{m_0 gz}{k_B T}\right) \tag{1.1.17}$$

where Laplace's barometric formula for ideal gas is used

$$\rho_0 = \rho_0^{(0)} \exp\left(-\frac{m_0 gz}{k_B T}\right) \tag{1.1.18}$$

Assuming $m_0 gz / k_B T < 1$ and expanding the formula (1.1.17) in series, we obtain (1.1.16) in linear approach. Thus, expression (1.1.16) is valid for the altitudes that fulfill the inequality:

$$\frac{m_0 gz}{k_B T} \leq 10^{-1} \tag{1.1.19}$$

where $T = 288,15^0 K$ is temperature on a sea level and we find, that $z \leq 850m$. Thus, the expression (1.1.16) is applicable in atmosphere only up to altitude of $850m$, or in water [7] where density practically does not depend on depth. \

The above arguments prove conclusively that the modern theory of sound waves in atmosphere of the Earth is incorrect and requires revision.

§ 2. Influences of heterogeneity of medium on the generation and propagation of sound

The neglect of the influence of gravity to on the generation and propagation of sound waves leads to contradiction in determining the speed of sound. As a matter of fact density of medium may be perturbed not only as a result of change mass in constant volume but by change of volume of constant mass of medium during temperature vibration under the conditions of constant pressure i.e. isobaricly. This mechanism of density perturbations in inhomogeneous medium cannot be neglected as entropy is changed when particles of medium are displaced which leads to alteration of temperature and thus to isobaric change of volume. Therefore sound may have not only adiabatic but isobaric speed as well. Combination of these two speeds must result in dependence of sound speed on altitude which constitutes qualitatively new result.

The urgency of the problem is exacerbated by the fact, that there are no accurate experimental measurements to dependence speed of sound on altitude. The paper [8] presents the results of measurement of sound speed at the see level for conditions and atm and it is shown that it precisely coincides with theoretical results. In the work [9] the results of measurements of dependence of sound speed on altitude in the Earth's atmosphere by means of a probe are given. There also demonstrated the chart of dependence of sound speed on the time of probe's vertical movement which does not enable us to determine precisely dependence of sound speed on altitude. From the above it is obvious that the correct determination of the velocity of sound in a heterogeneous environment is extremely important.

Density perturbation is transferred by the sound wave by means of mechanical displacement of particle of medium. According to established modern understanding density of atmospheric air, which represents ideal gas, depends only on pressure, i.e. $\rho = \rho(p)$. In paper [4] it is noted that such approach is justified only for homogeneous medium in every point of which entropy has the same meaning (isentropic medium). It is apparent that the Earth's atmosphere does not meet this condition and therefore $\rho = \rho(p,s)$. Consequently density perturbation should be put down as:

$$\rho' = \left(\frac{\partial \rho_0}{\partial P_0}\right)_s p' + \left(\frac{\partial \rho_0}{\partial S_0}\right)_p s' \qquad (1.2.1)$$

The first member in (1.2.1.) corresponds to density perturbation caused by alteration of mass in fixed volume as a result of pressure perturbation in conditions of constant entropy, while the second member corresponds to alteration of volume of fixed mass as a result of entropy perturbation in conditions of constant pressure. On the other hand, the disturbance entropy is caused by pressure perturbation, , i.e

$$s' = \left(\frac{\partial S_0}{\partial P_0}\right)_T p' \qquad (1.2.2)$$

(1.2.1) and (1.2.2) give

$$\rho' = \frac{1}{C^2} p' \qquad (1.2.3)$$

where

$$C^2 = \frac{C_s^2 C_p^2}{C_s^2 + C_p^2} \qquad (1.2.4)$$

$$C_s^2 = \left(\frac{\partial P_0}{\partial \rho_0}\right)_s \qquad (1.2.5)$$

$$C_P^2 = \left[\left(\frac{\partial \rho_0}{\partial S_0}\right)_P \left(\frac{\partial S_0}{\partial P_0}\right)_T\right]^{-1} \qquad (1.2.6)$$

If expression (1.2.5) is called adiabatic speed of sound, it is logical to call expression (1.2.6) isobaric speed of sound. Application the thermodynamic relations

$$\left(\frac{\partial \rho_0}{\partial S_0}\right)_P = \frac{T}{c_P}\left(\frac{\partial \rho_0}{\partial T}\right)_P \quad \text{and} \quad \left(\frac{\partial S_0}{\partial P_0}\right)_T = \frac{1}{\rho_0^2}\left(\frac{\partial \rho_0}{\partial T}\right)_P$$

for (1.2.6) results in

$$C_P^2 = \frac{c_P}{T}\left[\frac{\rho_0}{(\partial \rho_0/\partial T)_P}\right]^2 \qquad (1.2.7)$$

(1.2.7) demonstrates that in homogeneous medium square of sound speed coincides with coefficient which connects perturbations of pressure and density. We assume that the same is fair for heterogeneous medium and this coefficient should be defined by means of formula (1.2.4). We see that square of true value of sound speed equals to the reduced value of square of adiabatic and isobaric speeds. Such definition of sound speed - which is fair for any medium in the field of gravity, substantially change the existing concept of sound.

§ 3. Speed of Sound in Atmosphere of the Earth

Atmosphere of the Earth represents multilayered structure and in each layer, dependences of physical parameters on geometrical altitude z are different. On Pic. 1 which can be found in internet [2] as well as in scientific literature [1],[5] gives graphs which show that in interval of altitudes from $z = 0$ to $z \cong 11$ km (troposphere) temperature is reduced under strictly linear law $T = -6{,}52 \cdot 10^{-3} z + 288{,}15$. In the interval of altitudes from $z \equiv 11$ km to $z \cong 21$ km (tropopause) it is constant, in the interval from $z \cong 21$ km to $z \cong 51$ km (stratosphere) it increases and then up to z=85 km (mesosphere) again decreases approximately by law of $T = -2{,}60 \cdot 10^{-3}(z - 51 \cdot 10^3) + 270{,}50$. Such sharp deviation from dynamics of change of temperature in tropopause and stratosphere means that some anomalous process develops there as a result of which thermal energy is absorbed and thus medium is not adiabatic. Besides, with increase of altitude, experimental value of density is bigger than the values calculated by means of formula (1.1.18) . E.g. at the altitude $z = 7,5$ km where theoretical value of air density decreases e-times, relative error constitutes 23%, while at the altitude $z = 11$ km it constitutes 40%. At the altitude 35 km they equalize and thereafter theoretical value exceeds the experimental one and at the altitude of 85 km relative error constitutes 95%. In light of the aforementioned we assume that our theory can be applied up to 11 km altitude where fall of temperature strictly obeys of linear law and the adiabatic equation is fair. The aforesaid does not exclude possibility of its application in conditions of upper atmosphere.

Let's demonstrate that theoretical calculations which are shown above results in obvious dependence of sound speed on altitude in the Earth's atmosphere. Indeed, for value C_s we have expression (1.1.11) while by application of (1.1.18) from (1.2.7) for C_p we get

$$C_p = \sqrt{\frac{c_p k_B^2 T^3}{m_0^2 g^2 z^2}} = \left(\frac{c_p}{T}\right)^{1/2} \frac{k_B T^2}{m_0 g z} = \left(\frac{c_p}{T}\right)^{1/2} \frac{R T^2}{M g z} \quad (1.3.1)$$

Substituting (1.1.11) and (1.3.1) in (1.2.4) for expressions for speed of a sound in the specified layer ($0 \leq z \leq 11$ km) of the Earth atmosphere we receive

$$C = \sqrt{\frac{\gamma k_B T}{m_0 \left(1 + \gamma m_0 g^2 z^2 / c_p k_B T^2\right)}} = \sqrt{\frac{\gamma R T}{M \left(1 + \gamma M g^2 z^2 / c_p R T^2\right)}} \quad (1.3.2)$$

(1.3.2) shows that true value of sound speed in atmosphere obviously depends on altitude. Besides, we see that sound is adiabatic not only when $g = 0$ but when $g \neq 0$ and $z = 0$. At removal from a sea level, dependence of speed of sound on altitude becomes evident which is caused by heterogeneity of the atmosphere.

In Table 1, C_s, C_p and C values are given, who are obtained by formulas (1.1.11), (1.3.1) and (1.3.2) in the range of altitudes from to $z = 0$ up to $z = 11$ km, besides, corresponding values of speeds of sound C_{int} taken from the online calculator [2]. Values C_s and C_{int} coincide, and relative errors between values C and C_{int} at altitudes of $z = 1$ km and $z = 11$ km are equal to 0.3% and 33% respectively. It confirms our assumption that sound may be considered as adiabatic only up to altitude of $z = 850$ m. For visualization, dependences of C_s and C on z in intervals $(0-1)$ km and $(1-11)$ km are presented on the Figures 3 and 4 respectively.

Let's define the altitude on which $C_s = C_p$. From (1.1.11) and (1.3.1) it can be seen, that it is defined from a relation

$$z_0 = \sqrt{\frac{c_p k_B}{\gamma m_0} \frac{T}{g}} = \sqrt{\frac{c_p R}{\gamma M} \frac{T}{g}} \quad (1.3.3)$$

If we insert values of constant $c_p = 10^3$ J/kg °K in (1.3.3) and select values of temperature and acceleration of gravity from the tables provided for in monogtaph-guide [1] for altitude of

11

Table 1. Velocity values of C_s, C_p and C at altitudes from 0 to 11,000 m; comparison with internet data (online calculator).

z(m)	T°(K)	C_{int}	C_s	C_p	C
0	288.15	340.29	340.30	infinity	340.30
50.00	287.82	340.10	340.11	90180.02	340.11
100.00	287.50	339.91	339.92	45013.66	339.91
200.00	286.85	339.53	339.53	22430.54	339.50
300.00	286.20	399.14	339.15	14902.90	339.06
400.00	285.55	338.76	338.76	11139.12	338.61
500.00	284.90	338.38	338.38	8880.88	338.13
600.00	284.25	337.98	337.99	7375.42	337.64
700.00	283.60	337.60	337.61	6300.12	337.12
800.00	282.95	337.21	337.22	5493.66	336.59
900.00	282.30	336.82	336.83	4866.44	336.03
1000.00	281.65	336.43	336.44	4364.68	335.45
2000.00	275.15	332.53	332.54	2107.23	328.47
3000.00	268.65	328.58	328.59	1355.33	319.34
4000.00	262.15	324.58	324.59	979.83	308.12
5000.00	255.65	320.53	320.54	754.89	295.04
6000.00	249.15	316.43	316.44	605.24	280.42
7000.00	242.65	312.27	312.28	498.61	264.66
8000.00	236.15	308.06	308.07	418.87	248.18
9000.00	229.65	303.79	303.80	357.06	231.38
10000.00	223.15	299.46	299.47	307.81	214.65
10200.00	221.00	298.50	298.51	298.60	211.11
11000.00	216.65	295.07	295.08	267.69	198.26

$z = 10,2$ km ($T = 221^0 K$ and $g = 9,79$ m/sc^2), we will obtain $z_0 = 10,27$ km. For the same values of quantities formulae (1.1.11) and (1.3.1) give $C_s = 298,51$ m/sc, $C_p = 298,60$ m/sc. As we see the altitude calculated by means of formula (1.3.3) , at which adiabatic and isobaric speeds of sound are practically equal, is very close to the upper boundary of troposphere determined by the graph (Figure 1.) and to a high accuracy coincides with the altitude taken from the table which presupposes high reliability of our results. Thus, the upper boundary of troposphere is determined by the condition $C_s = C_p$. We assume that in case of fulfillment of this condition there occurs a resonance phenomenon activating absorption of external energy due to which temperature in tropopause remains constant while in stratosphere it increases. Experimental verification of this hypothesis can also be considered as a significant discovery. The results received in this paragraph are published in work [10].

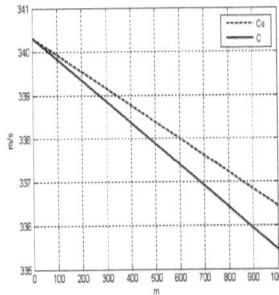

Fig3. C_s and C dependence on altitude from $z = 0$ to $z = 10^3$ m.

Fig3. C_s and C dependence on altitude from $z = 10^3$ to $z = 11 \times 10^3$ m.

§4. The coefficient of thermal expansion in the Earth's atmosphere

It was mentioned above that all thermodynamic characteristics of inhomogeneous medium must also be inhomogeneous, in other words they must depend on space coordinates. It was shown that this fundamental principle of physics is realized for the sound speed, which is the most important thermodynamic characteristic of medium. According to the modern theory the
sound speed is thought to be dependent only on temperature. Such understanding can be explained by the absence of experimental information concerning sound speed dependence on different altitudes, which was also mentioned above.

The situation is absolutely similar in terms of thermal expansion coefficient for ideal gases β
The first experiments that were supposed to determine this coefficient were conducted by G. Amontons (1661-1703) at the end of XVII century. Then this problem was studied by the outstanding scientists like Robert Boyle, Alessandro Volta, John Dalton, Joseph Priestley, Nicolas-Theodore de Saussure, Jacques Charles and others. The considerable contribution in the determination of coefficient was done by the great French scientist Joseph Luis Gay-Lussac (1778-1850). The results of experiments conducted by these scholars can be briefly defined as follows:

* At identical increasing of temperature, all the ideal gases are expanded identically.
* At constant pressure, volume of ideal gas linearly depends on temperature $t\,^\circ C$.

$$\upsilon = \upsilon_0(1+\beta t) \qquad (1.4.1)$$

where υ_0 is gas volume at $t=0°C$ and $\beta = 1/273.15\, grad^{-1}$ - is the coefficient of thermal expansion. After Kelvin had introduced absolute temperature scale $T(t)= (t+273,15)° K$, volume's dependence on temperature in the case of isobaric process looks like:

$$\upsilon = \upsilon_0 \beta T = \frac{\upsilon_0}{T_0} T \qquad (1.4.2)$$

and it is called Gay-Lussac's Law.

It is obvious that those experiments were conducted in the normal conditions, in other words, at a temperature of $T(0)=T_0=273,15° K$ and under a pressure of $P=1 atm.$. Experiments, for defining the dependence of coefficient of thermal expansion on altitude, had not been conducted. Consequently, according to Gay-Lussac's law $\upsilon_2 = \upsilon_1 T_2/T_1$ it is accepted that $\beta = 1/T_1$ at any altitude of the earth's atmosphere, where T_1 is the initial temperature of air.[11-13]. Such definition of β must be theoretically proved, so as coefficient of thermal expansion is characteristic of air to the same extent as sound speed is and for this reason in the Earth's atmosphere it must depend on the initial temperature T as well as on altitude z, i. e. $\beta = \beta(z,T)$. This article is concerned with the determination of this dependence.

Let's show that isobaric sound speed (1.2.7) can be expressed by β, for that let's transform the expression in square brackets

$$\left[\frac{\rho_0}{(\partial \rho_0/\partial T)_p}\right]^2 = \left[\frac{m}{\upsilon}\left(\frac{\partial(m/\upsilon)}{\partial T}\right)_p^{-1}\right]_{m=const}^2 = \left[\frac{1}{\upsilon}\left(\frac{\partial \upsilon}{\partial T}\right)_p\right]^{-2} = \frac{1}{\beta^2} \qquad (1.4.3)$$

where $\beta = (1/\upsilon)(\partial \upsilon/\partial T)_p$ - is the coefficient of thermal expansion. Thus we obtain for C_p

$$C_p = \frac{1}{\beta}\left(\frac{c_p}{T}\right)^{1/2} \qquad (1.4.4)$$

If we compare the formula (1.4.4) with the formula (1.3.1) for the coefficient of air expansion we can find

$$\beta = \frac{Mgz}{RT^2} = \frac{Mgz}{RT}\frac{1}{T} \qquad (1.4.5)$$

From (1.4.5) it is clear that actually $\beta = \beta(z,T)$. But on the other hand $\beta = 0$ if $z=0$, i. e. on the sea level when the air is heated, it is not expanded. The absurdity of this result is obvious, but it doesn't mean that the formula (1.4.5) has no physical meaning. In order to prove this let's calculate changes of β in the altitude interval $0 \leq z \leq z_0$ where

$z_0 = 10{,}27\ km$ is the troposphere's upper boarder, which is determined by the condition $C_s = C_p$ and within this boundary our theory is true. According to the chart from the monograph-guide [1] we find $T(z_0) = 221^\circ K$ and then

$$\Delta\beta = \beta(z_0) - \beta(0) = Mgz_0/RT^2(z_0) = (1/T(z_0))(c_p M/\gamma R)^{1/2} = 7{,}1\times 10^{-3}\,(^\circ K)^{-1}$$

The change of β on the 1m. altitude will be $\Delta\beta/z_0 = 7\times 10^{-7}\,(^\circ Km)^{-1}$. Thus, β depends on altitude very weakly and that's why it can be averaged over z. In order to do it let's express z_0 and $\beta(z,T)$ through the adiabatic index. Using known relation $c_p - c_v = R/M \Rightarrow \gamma - 1 = R/c_v M$ from the formulas (1.4.3) and (1.5.3) it is easy to find:

$$z_0 = \frac{c_v T \sqrt{\gamma - 1}}{g} \qquad (1.4.6)$$

$$\beta(z,T) = \frac{gz}{c_v T^2 (\gamma - 1)} \qquad (1.4.7)$$

Let's now average $\beta(z,T)$ in the altitude interval $0 < z \le z_1$ where $z_1 = \varepsilon z_0$ and ε is dimensionless parameter determining maximal altitude in the Earth's atmosphere for which Gay-Lussac's Law runs, i. e.

$$z_1 = \varepsilon z_0 = \frac{\varepsilon c_v T \sqrt{\gamma - 1}}{g} \qquad (1.4.8)$$

Simple calculations for average meaning of $\overline{\beta}(T)\big|_0^{z_1}$ give us:

$$\overline{\beta}(T)\big|_0^{z_1} = \frac{1}{z_1}\int_0^{z_1}\beta(z,T)dz = \frac{0{,}5\varepsilon}{\sqrt{\gamma - 1}}\frac{1}{T} \qquad (1.4.9)$$

According to Gay-Lussac's Law it must be demanded:

$$\overline{\beta}(T)\big|_0^{z_1} = \frac{1}{T} \Rightarrow \varepsilon = 2\sqrt{\gamma - 1} \qquad (1.4.10)$$

The expression (1.4.10) has an important physical meaning, for it shows that for different gases a parameter ε and correspondingly the altitude z_1, above which Gay-Lussac's Law doesn't work, have different meanings

$$z_1(\gamma, c_v, T) = 2\sqrt{\gamma - 1}\, z_0 = \frac{2(\gamma - 1)c_v T}{g} \qquad (1.4.11)$$

In the table 2, there are presented values of z_0 and z_1 for different gases according to growth of their molar mass, notably z_0 for every gas denotes altitude where $C_s = C_p$ if the Earth's atmosphere was composed only of that gas. We'd like to note that for air the temperature value was taken according to altitude from the table of the monograph-guide [1] As there is no such table for other gases, instead of T in the formulas (1.4.6) and (1.4.11) we take its average value all over the troposphere and according to the same chart it equals $\overline{T} \cong 254^\circ K$.

Table 2. Dependence z_0 and z_1 on adiabatic index and molar mass M.

Gas or Vapor	Formula	Molar Mass M g/mol	Specific Heat C_V kJ/kg$^\circ$K	Ratio of Specific Heats γ	Z_0 km	Z_1 km
Hydrogen	H_2	2	10,16	1,41	168,6	215,9
Helium	He	4	3,12	1,66	66,3	107,7
Ammonia	NH_3	17	1,66	1,32	23.6	27.5
Hydroxyl	OH	17	1,27	1,41	21.1	27,0
Water vapor	H_2O	18	1,46	1,32	21.4	24,2
Neon	Ne	20	0,62	1.66	13,0	21,1
Acetylene	C_2H_2	26	1,37	1,23	17,0	16,3
Nitrogen	N_2	28	0,74	1,40	12,1	15,3
Carbon monoxide	CO	28	0,72	1,40	11,8	14,9
Air		29	0,72	1,40	11,9	14,9
Nitric oxide	NO	30	0,72	1,38	11,4	14,1
Oxygen	O_2	32	0,66	1,39	10,6	13,3
Argon	Ar	40	0,31	1,67	6,6	10,8
Propylene	C_3H_6	42	1,31	1,15	13,2	10,2
Propane	C_3H_8	44	1,48	1,13	13,9	10,0
Carbon dioxide	CO_2	44	0,65	1,29	9,1	9,8
Nitrous oxide	N_2O	44	0,69	1,27	9,3	9,7
Butane	C_4H_{10}	58	1,53	1,09	11.8	7,1
Sulfur dioxide	SO_2	64	0,51	1,25	6,6	6,6
Chlorine	Cl_2	71	0,36	1,33	5,4	6,2
Xenon	Xe	131	0,10	1,65	2,0	3,3

Analysis of table 2. permits us to make the following conclusions:
1) For the gases whose number of atoms $N \leq 3$, the inequality $z_0 < z_1$ is fulfilled. It means that

these gases follow Gay-Lussac's Law all over the interval $0 \leq z \leq z_1$. According to above offered hypothesis in the altitude interval $z_0 < z \leq z_1$ medium adiabaticity is violated altogether and our theory is not true.

2) For the gases whose number of atoms $N > 3$ the inequality $z_0 > z_1$ is fulfilled. It means that these gases follow Gay-Lussac's Law only in the interval $0 \leq z \leq z_1$. In the altitude interval $z_1 < z \leq z_0$ medium is adiabatic, but Gay-Lussac's Law not performed.

3) As the molar mass of gas increases, altitude z_1 (below of which Gay-Lussac's Law is fulfilled) decreases. It can by explained by the fact that the more is the molar mass of gas the more is the influence of the Earth's gravitational field and properties of gas are more removed from the ideal one.

It is interesting to determine an average value of thermal expansion coefficient for the case 2) in the interval $z_1 < z \leq z_0$. The simple calculation by means of the formulas (1.4.6), (1.4.7) and (1.4.11) give us:

$$\overline{\beta}(T)\Big|_{z_1}^{z_0} = \frac{1}{z_0 - z_1} \int_{z_1}^{z_0} \beta(z,T)dz = \left(\frac{1}{2\sqrt{\gamma-1}} + 1\right)\frac{1}{T} \qquad (1.4.12)$$

As we can see, the influence of the gravitational field of polyatomic gases $(N > 3)$ is revealed not only in the fact that $z_1 < z_0$, but also in the fact that in the altitude range $z_1 < z \leq z_0$ their coefficient of thermal expansion is more than usual by value $1/(2T\sqrt{\gamma-1})$.

Thus, the commonly held opinion on universality of the laws of ideal gas that has existed in science for more than two centuries is erroneous. These laws were obtained by means of experiments carried out at the sea level where the Earth's atmosphere can be considered as homogeneous medium [10],[14]. Influence of Earth's gravitational field on the thermodynamic parameters of ideal gases, depending on their molar mass and the adiabatic index , becomes more evident with increase of altitude.

Our results are based on the assumption that all gases discussed in this work at sufficient approximation can be considered as ideal ones. Proceeding from this consideration we think, that distribution of densities of all gases in the Earth's gravitational field obeys Laplasse's barometric formula (1.1.18) and the upper border, up to which medium's adiabaticity is retained, can be defined from the condition $C_s = C_p$.

We have shown, that for the air this altitude coincides with the upper border of troposphere $(z_0 = 10,27 km)$ and at temperature in this altitude, which is taken from the graph given in the monograph-guide [1] condition $C_s = C_p$ performed with high precision. It testifies the reliability of our theory and its results. For different gases this altitude will be different, but they have in common the fact that up to this altitude adiabatic speed of sound in them is less than isobaric speed $(C_s < C_p)$, in other words, the compressibility of medium prevails over. its incompressibility [14].

At higher altitude z_0 where our theory is not true, we suppose that a inequality sign between C_s and C_p changes into an opposite $(C_s > C_p)$ and consequently, the change of medium properties from compressibility to incompressibility takes place. Thus, altitude z_0 in the Earth's gravitational field is like the Rubicon –by crossing of it, the properties of medium are changing into the opposite. On this bases we think that at this altitude, when $C_s = C_p$, some anomalous, resonant processes occur and they are connected with absorption of thermal energy (for example, infrared solar radiation). It is clearly visible by the example of tropopause (fig.1), where fall of the air temperature suddenly stops and during the whole tropopause remains constant while in the stratosphere it increases.

Results

The reasoning and estimations given in this chapter vividly show the importance of existing problems in modern gasdynamics. Perhaps these very problems are the reason why the upper layers of atmosphere have not been examined sufficiently. It goes without saying that we don't profess to eradicate all these problems, however we suppose, that we have laid foundation for qualitatively new explorations in this important field of physics. The most important results received in this chapter are the following:

- It is shown that the existing theory of sound waves, which presupposes the presence of only one mechanism of generation and distribution of sound and is applicable for the whole atmosphere, is true only for homogenous medium.
- Atmosphere can be regarded as homogeneous till the altitude $z \leq 850m$. Above this altitude the influence of gravitational field of Earth starts to evolve.
- In the inhomogeneous medium together with adiabatic mechanism there is also isobaric mechanism of generation and propagation of sound.
- Sound possesses two speeds of propagation; adiabatic $C_s(T)$ and isobaric ones $C_p(z,T)$
- The true value of squared velocity of sound is reduced from square these two speeds, which leads to its obvious dependence on altitude z.
- The upper border of troposphere, up to which atmosphere can be regarded as an adiabatic medium, is determined with high precision from the equation $C_s(T) = C_p(z,T)$.
- The thermal expansion coefficient of an ideal gas in the atmosphere of the Earth depends on the height $\beta = \beta(z,T)$.
- In the atmosphere of Earth the altitude up to which laws of ideal gases are applicable,, decreases with the increase of molar mass of gas.

The results received in this chapter are published in the works [2] and [15].

CHAPTER II

Compressibility or incompressibility as properties of an inhomogeneous medium

The Problems

The analysis of present research literature shows, that the most problems of gas and hydrodynamics are solved for the incompressible model of medium. Such approach is caused by two reasons: first, consideration of compressibility makes difficult to solve of system of hidrodynamic equations and secondly, it is not connected with the property of the medium itself, but it is conceived as some approximation, which follows from the conditions of the problem. The medium is considered as incompressible ($C_s = \infty$) if $C_s \gg V_0$, where V_0 is a characteristic speed of liquid motion. From this it follows that in the course of investigation of wave processes (except sound wave), any stationary medium can be considered as incompressible. Such approach leads to falling out the equation of state of medium $\rho' = p'/C_t^2$, which connecting perturbed values of density and pressure, from the system of linearized hydrodynamic equations. The essence of this equation is also that it contains thermodynamic parameters of medium and in case of its absence it is unclear, which medium is considered. Besides, it is obvious that in this case perturbation of density does not occur ($\rho' = 0$) and distribution of wave is impossible. That is why many gas and hydrodynamic tasks like the tasks of surface gravity wave and internal gravity wave are solved incorrectly.

§ 1. Internal gravity waves, as an example of the incorrectness of modern hydrodynamics

Let's look how is solving the problem of internal gravity waves in the monograph [5]. Based on the incompressibility of medium they neglect change of density related to change of pressure and assume that change of density can only be isobaric, at the expense of change of entropy under mechanic oscillation of inhomogeneous medium, i.e.

$$\rho' = \left(\frac{\partial \rho_0}{\partial s_0}\right)_p s' \qquad (2.1.1)$$

Thereafter they write down linearized equations of motion (1.1.1) and mass continuity (1.1.6) in the form

$$\frac{\partial \vec{v}}{\partial t} = \frac{\vec{g}}{\rho_0}\left(\frac{\partial \rho_0}{\partial s_0}\right)_p s' - \nabla \frac{p'}{\rho_0} \qquad (2.1.2)$$

$$\nabla \vec{V} = 0 \qquad (2.1.3)$$

Having presented all perturbed variable values in the form of $f'(\vec{r},t) = const \times \exp[i(\vec{k}\vec{r} - \omega t)]$, from equations (1.1.3), (2.1.2) and (2.1.3) they obtain

$$i\omega s' = \vec{V}\nabla s_0 \qquad (2.1.4)$$

$$-i\omega \vec{V} = \frac{1}{\rho_0}\left(\frac{\partial \rho_0}{\partial s_0}\right)_p s'\vec{g} - \frac{i\vec{k}}{\rho_0}p' \qquad (2.1.5)$$

$$\vec{k}\vec{V} = 0 \qquad (2.1.6)$$

Multiplying the equation (2.1.5) by vector $\vec{k}$ they receive

$$ik^2 p' = \left(\frac{\partial \rho_0}{\partial s_0}\right)_p s'(\vec{g}\vec{k}) \qquad (2.1.7)$$

From equations (2.1.4), (2.1.5) and (2.1.7) can easily be obtained dispersive equation in the form of

$$\omega^2 = \omega_0^2 \sin^2 \theta \qquad (2.1.8)$$

where

$$\omega_0^2 = -\frac{g}{\rho_0}\left(\frac{\partial \rho_0}{\partial s_0}\right)_p \frac{ds_0}{dz} \qquad (2.1.9)$$

and θ is angle between vector $\vec{k}$ and axis Z. Thus, we get some kind of strange transverse wave, frequency of which depends only on direction of wave vector and it can be of any length. Below we show that the reason for this paradox is the incorrect understanding of the criterion incompressibility of medium.

§ 2. Equation of mass continuity for inhomogeneous medium

The Equation (1.1.2) determines change of density liquid particle in given volume υ and says that it equals to difference of mass flows incoming and outgoing through the surface which limits this volume However, change of density may occur as at the cost of change of substance mass in constant volume, also by change of volume of constant mass of substance. Indeed:

$$\frac{d\rho}{dt} = \frac{d}{dt}\left(\frac{m}{\upsilon}\right) = \frac{\upsilon \frac{dm}{dt} - m \frac{d\upsilon}{dt}}{\upsilon^2} = \left(\frac{d\rho}{dt}\right)_\upsilon - \left(\rho \frac{d}{dt}\ln \upsilon\right)_m \qquad (2.2.1)$$

Assuming in second summand of the right part of the Equation (2.2,1) $m=1$ i.e.. $\rho = 1/\upsilon$, we obtain:

$$\left(\rho \frac{d}{dt}\ln \upsilon\right)_m = -\left(\rho \frac{d}{dt}\ln \rho\right)_m = -\left(\frac{d\rho}{dt}\right)_m$$

and finally from (2.2.1) we find out

$$\frac{d\rho}{dt} = \left(\frac{d\rho}{dt}\right)_\upsilon + \left(\frac{d\rho}{dt}\right)_m \qquad (2.2.2)$$

Hence, complete change of density consists of two parts, the first of which is determined by the equation (1.2.2), i.e.

$$\left(\frac{d\rho}{dt}\right)_v = \frac{\partial \rho}{\partial t} = -div(\rho \vec{V}) \qquad (2.2.3)$$

The second part describes change of density of the substance of constant mass as a result of change of volume which may occur only due to change of temperature which, in turn, in the absence of heat source, is possible only under change of entropy, i.e.

$$\left(\frac{d\rho}{dt}\right)_m = \left(\frac{\partial \rho_0}{\partial S_0}\right)_p \frac{\partial s}{\partial t} = -\left(\frac{\partial \rho_0}{\partial S_0}\right)_p (\vec{V}\nabla)s \qquad (2.2.4)$$

Adiabatic equation (1.1.3) is used here. For homogeneous (isentropic) medium $(\vec{V}\nabla)s = 0$ and (2.2.2) makes obvious that change of density of liquid particle is indeed determined under the equation (1.2.2). For non-homogeneous medium it can be written

$$(\vec{V}\nabla)s = \vec{V}\left(\frac{\partial S_0}{\partial P_0}\right)_T \nabla p \qquad (2.2.5)$$

and then, proceeding from (2.2.2-5) the equation of continuity of mass in inhomogeneous media will have the form of:

$$\frac{d\rho}{dt} = -\rho \nabla \vec{V} - \frac{\vec{V}\nabla p}{C_p^2} \qquad (2.2.6)$$

where

$$\frac{d\rho}{dt} = \frac{\partial \rho}{\partial t} + (\vec{V}\nabla)\rho \qquad (2.2.7)$$

and the value C_p^2 have the dimension of speed square and equals to

$$C_p^2 = \left[\left(\frac{\partial \rho_0}{\partial S_0}\right)_p \left(\frac{\partial S_0}{\partial P_0}\right)_T\right]^{-1} \qquad (2.2.8)$$

which coincides with the formula (1.2.6).

Thus, equation (2.2.6) make apparent that in inhomogeneous medium, even though it is incompressible, its density all the same can change because of perturbation of entropy. In light of the aforementioned, the equation of continuity of mass for inhomogeneous media should have the form:

$$\frac{d\rho}{dt} = \frac{\partial \rho}{\partial t} + (\vec{V}\nabla)\rho = -\rho\nabla\vec{V} - \frac{\vec{V}\nabla p}{C_p^2} \qquad (2.2.9)$$

Exactly this is the generalized equation of mass continuity which for homogeneous medium $(C_p = \infty)$ grades into the equation which was considered as universal and is applied in the contemporary theory for any medium. Thus, compressibility of homogeneous medium cannot be ignored since from (2.2.9) it follows that in such case $(\nabla\vec{V} = 0)$, change of density and generation of wave processes is impossible.

Now the reason of paradox associated with internal gravity waves is clear: if in the equation of mass continuity $d\rho/dt = -\rho\nabla\vec{V} - \vec{V}\nabla p/C_p^2$, the second summand related to heterogeneity of medium is not taken into consideration, then the condition $\nabla\vec{V} = 0$ (incompressibility medium) means that $\rho' = 0$ and following from (2.1.1) $s' = 0$ and then all equations from which dispersive equation (2.1.8) is obtained are nulling.

Our reasoning conclusively proves that there is no ideally homogeneous medium in nature. Any medium is inhomogeneous to a greater or lesser degree depending on correlation C_s and C_p. Moreover, homogeneity of medium does not mean that it is not influenced by external field of force. For instance, in the Earth atmosphere isobaric speed of sound is determined by formula (1.3.1) from which it follows that at sea level $(z = 0)$ $C_p = \infty$, which is the criterion of ideally homogeneous medium, however, the formula includes acceleration of gravity g. Thus, only strongly inhomogeneous medium when $C_s \gg C_p$, can be incompressible, while weakly inhomogeneous medium when $C_p \gg C_s$ can be compressible. Hence, the criterion of compressibility or incompressibility is determined not from correlation of C and V_0, but from correlation of C_s and C_p, and the wide-spread opinion that compressibility (incompressibility) and homogeneity (heterogeneity) are independent concepts is erroneous. The reason of incorrectness discovered by us in the existing theory of internal gravity waves is that incompressibility of medium does not mean that $\nabla\vec{V} = 0$ or $C_s = \infty$, which is the same. It is also necessary to take into consideration existence of isobaric sound speed C_p, which represents the measure of heterogeneity of medium and is present in the equation of mass continuity.

It is interesting to apply the formula (1.4.4) for condensed mediums. For example, for water $\beta = 1.5 \times 10^{-4}\,^0 K^{-1}$, $c_p = 4.19 \times 10^3\,\text{J/kg}\,^0 K$ and then from formula (1.4.4) at temperature $T = 288^0 K$, we will get $C_p = 25210\,\text{m/sec}$. On the other hand, according to the experiment, sound speed in water at the same temperature to a high accuracy equals to $C = 1480\,\text{m/sec}$. Then from formula (1.2.4) we have $C_s = C\,C_p/\sqrt{C_p^2 - C^2} = 1482.6\,\text{m/sec}$. As we see, in water, sound speed practically equals to adiabatic sound speed i.e. $C = C_s$. Similar calculations, for example, for iron ($\beta = 33.9 \times 10^{-6\,0} K^{-1}$, $c_p = 476.4\,\text{J/kg}\,^0 K$, $T = 288^0 K$, $C_p = 38968\,\text{m/sec}$, $C = 5130$ m/sec) give $C_s = 5175\,\text{m/sec}$, and sound speed in iron is also adiabatic $(C = C_s)$. Consequently, according to the new determination of criteria of compressibility and incompressibility, water and iron should be considered as compressible media in terms of thermodynamics and incompressibility condition $\nabla \vec{V} = 0$ must not be applied to them. Moreover, it may seem paradoxical, but water and iron are more compressible than air in upper layers of the atmosphere. In fact, for example, at a height of 10-11 km, $C_s \approx C_p \approx 300$ m/sec. Thus, the terms "compressibility" and "incompressibility" in thermodynamics characterize not the state of aggregation of matter, but physical process of sound propagation in it.

§ 3. Changing of densities of the flows of some of the physical quantities in inhomogeneous medium

Discourses given in the previous paragraph play an important part in determining the changes of some physical values in the moving non-homogenous mediums. Let us consider for example the change of liquid mass in the predefined motionless volume, when liquid moves at a speed of $\vec{V}$. In order to get that done, let us write the equation (2.2.9) by the following form:

$$\frac{\partial \rho}{\partial t} = -div(\rho \vec{V}) - \frac{\vec{V}\nabla p}{C_p^2} \qquad (2.3.1)$$

and integrate over given volume, and then apply the Gauss-Ostrogradski theorem. After this we will get:

$$\frac{\partial m}{\partial t} = -\oint \rho \vec{V} d\vec{f} - \frac{1}{C_p^2}\int \vec{V}\nabla p d\upsilon \qquad (2.3.2)$$

where m is the liquid mass in the given volume and $d\vec{f}$ is the vector whose magnitude equals to element area of surface, which limiting given volume and directed along outer normal. In strongly inhomogenous medium a surface integral in (2.3.2) is identically equal zero even when $V \neq 0$ (incompressible medium), though a volumetric integral vanishes only if $V = 0$. Consequently, in the moving highly inhomogenous medium the word-combination "constant mass in a fixed volume" is meaningless.

Let us consider change of density of energy flow of immovable volume υ of liquid, i.e. calculate integral

$$\frac{\partial}{\partial t}\int_\upsilon \left(\frac{\rho V^2}{2} + \rho\varepsilon\right) d\upsilon \qquad (2.3.3)$$

where ε is inner energy of unit mass of liquid. Let's first calculate the first summand of (2.3.3)

$$\frac{\partial}{\partial t}\left(\frac{\rho V^2}{2}\right) = \frac{V^2}{2}\frac{\partial \rho}{\partial t} + \rho \vec{V} \frac{\partial \vec{V}}{\partial t} \qquad (2.3.4)$$

Having applied the generalized equation of mass continuity (2.3.1) and the motion equation of Euler

$$\rho \frac{\partial \vec{V}}{\partial t} = -\nabla p + \rho \vec{g} - \rho(\vec{V}\nabla)\vec{V} \qquad (2.3.5)$$

from (2.3.4) we will obtain

$$\frac{\partial}{\partial t}\left(\frac{\rho V^2}{2}\right) = -\frac{V^2}{2}\left(\frac{\vec{V}\nabla p}{C_p^2} + div\rho\vec{V}\right) + \rho\vec{V}\vec{g} - \frac{\rho\vec{V}}{2}\nabla V^2 - \vec{V}\nabla p \qquad (2.3.6)$$

Here the correlation $\vec{V}(\vec{V}\nabla)\vec{V} = \vec{V}\nabla V^2/2$ is used. From the definition of differential of thermal function of unit mass of liquid $dw = Tds + dP/\rho$, we can write

$$\nabla P = \rho \nabla w - \rho T \nabla s \qquad (2.3.7)$$

And then, from formula (2.3.6) we will obtain

$$\frac{\partial}{\partial t}\left(\frac{\rho V^2}{2}\right) = -\frac{V^2}{2}\left(\frac{\vec{V}\nabla p}{C_p^2} + div\rho\vec{V}\right) + \rho\vec{V}\vec{g} - \rho\vec{V}\nabla\left(w + \frac{V^2}{2}\right) + \rho T\vec{V}\nabla s \qquad (2.3.8)$$

Using the correlation $d\varepsilon = Tds + (p/\rho^2)d\rho$, and equation (1.1.3) from the second summand of (2.3.3), we will find out

$$\frac{\partial}{\partial t}(\rho\varepsilon) = \left(\varepsilon + \frac{p}{\rho}\right)\frac{\partial \rho}{\partial t} + \rho T \frac{\partial s}{\partial t} = -\left(\varepsilon + \frac{p}{\rho}\right)\left(div\rho\vec{V} + \frac{\vec{V}\nabla p}{C_p^2}\right) - \rho T\vec{V}\nabla s \qquad (2.3.9)$$

Applying (2.3.8) and (2.3.9), in formula (2.3.3) we will obtain

$$\frac{\partial}{\partial t}\int\left(\frac{\rho V^2}{2} + \rho\varepsilon\right)dv = -\int div\left[\rho\vec{V}\left(\varepsilon + \frac{p}{\rho} + \frac{V^2}{2}\right)\right]dv - \\ \frac{1}{C_p^2}\int\frac{\nabla p}{\rho}\left[\rho\vec{V}\left(\varepsilon + \frac{p}{\rho} + \frac{V^2}{2}\right)\right]dv + \int\rho\vec{V}\vec{g}dv \qquad (2.3.10)$$

Having reconstructed the first volumetric integral in the right part of equation (2.3.10) into superficial, we will ultimately obtain

$$\frac{\partial}{\partial t}\int\left(\frac{\rho V^2}{2}+\rho\varepsilon\right)dv = -\oint\rho\vec{V}\left(\varepsilon+\frac{V^2}{2}\right)d\vec{f} - \oint p\vec{V}d\vec{f} -$$
$$\frac{1}{C_p^2}\int\nabla p\vec{V}\left(\varepsilon+\frac{v^2}{2}+\frac{p}{\rho}\right)dv + \int\rho\vec{V}\vec{g}dv \quad (2.3.11)$$

Change of density of flow of impulse in immovable volume of liquid is described by derivative $\partial(\rho\vec{V})/\partial t$. Let us consider this derivative in components, i.e.

$$\frac{\partial(\rho V_i)}{\partial t} = \rho\frac{\partial V_i}{\partial t} + V_i\frac{\partial\rho}{\partial t} \quad (2.3.12)$$

and write equations (2.2.9) and (2.3.5) in the forms

$$\frac{\partial\rho}{\partial t} = -\frac{\partial(\rho V_k)}{\partial x_k} - \frac{V_k}{C_p^2}\frac{\partial p}{\partial x_k} \quad (2.3.13)$$

$$\rho\frac{\partial V_i}{\partial t} = -\frac{\partial p}{\partial x_k}\delta_{ik} - \rho g n_i - \rho V_k\frac{\partial V_i}{\partial x_k} \quad (2.3.14)$$

where indexes $k; i = 1;2;3$, moreover $(x_1; x_2; x_3) \to (x; y; z)$ and n_i is the components of unit vector directed along the axis Z $(\vec{n} = \vec{e}_z)$. Having substituted (2.3.13) and (2.3.14) in (2.3.12) we will obtain

$$\frac{\partial(\rho V_i)}{\partial t} = -\frac{\partial}{\partial x_k}(p\delta_{ik} + \rho V_i V_k) - \rho g n_i - \frac{V_i V_k}{C_p^2}\frac{\partial p}{\partial x_k} \quad (2.3.15)$$

Integrating equation (2.3.15) over given volume v, gives the following:

$$\frac{\partial}{\partial t}\int\rho V_i dV = -\oint(p\delta_{ik} + \rho V_i V_k)df_k - mgn_i - \frac{1}{C_p^2}\int v_i v_k\frac{\partial p}{\partial x_k}dV \quad (2.3.16)$$

Correlations (2.3.11) and (2.3.16) for homogeneous liquids are obtained in the monograph [5], where only superficial integrals are present. Consequently, change of densities of flows of energy and impulse in homogeneous liquids is influenced only by superficial processes. In inhomogeneous media, speeds of change densities flows of energy and impulse are depend also on voluminous processes conditioned by work of external force and natural change of pressure inside the given volume.

Results

Thus in this chapter we have shown:

- Application of criterion of incompressibility to any medium, contradicts to the real meaning of this term.
- On the basis of expression of speed of sound in inhomogeneous medium and generalized equation of continuity of mass, it is proved that so called internal gravitation waves do not exist in nature. This mistake is caused by incorrect determination of incompressibility of medium.
- Correct understanding of criteria of compressibility or incompressibility leads to qualitatively new understanding of homogeneity or heterogeneity of medium, in particular - only strongly inhomogeneous medium $(C_s \gg C_p)$ can be incompressible while weakly inhomogeneous medium $(C_s \ll C_p)$ is always compressible.
- In inhomogeneous medium additional terms are added to known hydrodynamic (gas dynamic) correlations which disappear while transferring to homogeneous model of medium.
- It is shown that in the thermodynamic sense condensed media is more compressible than the rarefied gases.

These results are published in work [14].

CHAPTER III

Peculiarities of sound waves in inhomogeneous medium

The Problems

It was shown above that the disregard of the influence of gravitational field leads to the incorrect perception of such fundamental properties of medium like compressibility and incompressibility, as well as to the incorrect solution of many important applied problems. One of them is generation and propagation of sound waves in the inhomogeneous medium. Solving these problems is impossible without taking into account the influence of the gravitational field, since many of the effects associated with this influence, such as the dependence of the sound velocity in the Earth's atmosphere on the height, the change in wavelength in the propagation of sound in the vertical direction, the refraction of waves in the propagation of sound in the horizontal direction and others, remain out of sight of the researcher. In modern theory of sound waves it is believed that the Earth's gravitational field does not affect the generation and propapagation of sound waves. Below it will be shown that the equation of sound wave is a special case of generalized equation of gravitational waves in the medium, and thus, the conventional wisdom that the sound wave is not a gravitational wave is not correct.

§ 1. Generalized equation of gravitational waves in moving inhomogeneous medium

Above it was shown that the expression of adiabatic speed of sound $C_s = (\gamma kT/m_0)^{1/2}$ with the help of which the speed of sound is currently determined as in lower as well as upper layers of the atmosphere, is applicable only up to the altitudes $z = 10^3$ m. Above this altitude the obvious dependence of the sound speed on coordinates z is revealed, along which the atmosphere is inhomogeneous, as a result of influence of gravitational field of the Earth. Consideration of influence of gravitational field became possible after entropy had been included into the equation of the state of the atmosphere which is considered as an ideal gas.

We wrote this equation in the form of $\rho = \rho(p,s)$ instead of generally accepted $\rho = \rho(p)$ which is true only for isentropic media and is not applicable to the Earth's atmosphere. Such approach enabled us to conclude that along with adiabatic mechanism of generation of sound wave, there exists an isobaric one and exactly this mechanism leads to dependence of the speed of sound on altitude z or on density, which is the same. The true value of the square of sound speed, which is reduced from the squares of adiabatic and isobaric speeds of sound $\left(C^2 = C_s^2 C_p^2 / \left(C_s^2 + C_p^2\right)\right)$, is obtained from equation of the state of the medium and is defined as the coefficient which connecting perturbations of pressure and density $(p' = C^2 \rho')$.

Here we will show that the equation of the sound wave is obtained from the equation of gravitational wave in inhomogeneous medium under certain conditions, and thus, the sound wave is a special case of a gravitational wave. Gas-dynamic (hydrodynamic) system of equations in the adiabatic non-dissipative inhomogeneous medium has the form

$$\begin{cases} \rho\left[\dfrac{\partial \vec{V}}{\partial t} + (\vec{V}\nabla)\vec{V}\right] = -\nabla p + \rho \vec{g} \\ \dfrac{\partial \rho}{\partial t} + (\vec{V}\nabla)\rho = -\rho \nabla \vec{V} - \dfrac{\vec{V}\nabla p}{C_p^2} \end{cases} \quad (3.1.1)$$

Here adiabatic equation (2.2.5) has already been used. Let's assume that the medium is moving with constant speed $\vec{V}_0$ and let's present all its variable members in the form of the sum of their stationary and disturbed values, i.e.

$$\vec{V} = \vec{V}_0 + \vec{V}', \quad p = P_0 + p', \quad \rho = \rho_0 + \rho' \quad (3.1.2)$$

Let us consider the linear wave, when $\vec{V}'$, P', , and ρ', and their derivatives to be small quantities and let's neglect members containing their squares. Then, the linearized system of equations (3.1.1) will take the form of

$$\begin{cases} \rho_0\left[\dfrac{\partial \vec{V}}{\partial t} + (\vec{V}_0 \nabla)\vec{V}\right] = -\nabla p + \dfrac{\vec{g}}{\tilde{C}^2} p \\ \dfrac{1}{\tilde{C}^2}\dfrac{\partial p}{\partial t} + (\vec{V}_0 + \vec{V})\nabla \rho_0 + \dfrac{\vec{V}_0}{\tilde{C}^2}\nabla p = -\rho_0 \nabla \vec{V} - \dfrac{(\vec{V}_0 + \vec{V})\rho_0 \vec{g} + \vec{V}_0 \nabla p}{\tilde{C}_p^2} \end{cases} \quad (3.1.3)$$

The following equations are used here:
Condition of equilibrium of fluid in gravitational field of the Earth

$$\nabla P_0 = \rho_0 \vec{g} \quad (3.1.4)$$

The equation of state of the liquid

$$\rho = \frac{p}{\tilde{C}^2} \tag{3.1.5}$$

and marks of disturbed values are omitted. $\tilde{C}^2$ is the generalized meaning of squared velocity of sound in moving medium $\tilde{C}^2 = (C - V_0)^2$, which is reduced from squares of adiabatic $\tilde{C}_s = C_s - V_0$ and isobaric $\tilde{C}_p = C_p - V_0$ speeds of sound in respect of the moving medium

$$\tilde{C}^2 = \frac{\tilde{C}_s^2 \tilde{C}_p^2}{\tilde{C}_s^2 + \tilde{C}_p^2} \tag{3.1.6}$$

Expressions C_s and C_p determined by the formulas (1.1.11) and (1.2.7) are relevant speeds of sound in respect of immovable medium and they are characterizing degree of its homogeneity or inhomogeneity.

Let's apply operator ∇ to the first equation (3.1.3) and operator $\partial/\partial t$ to the second one and then we will get:

$$\begin{cases} \nabla\left(\rho_0 \dfrac{\partial \vec{V}}{\partial t}\right) + \Delta p - \vec{g}\,\dfrac{\tilde{C}^2 \nabla p - 2\tilde{C} p \nabla \tilde{C}}{\tilde{C}^4} = -\nabla\left[\rho_0 (\vec{V}_0 \nabla)\vec{V}\right] \\ \dfrac{1}{\tilde{C}^2}\dfrac{\partial^2 p}{\partial t^2} + \nabla\left(\rho_0 \dfrac{\partial \vec{V}}{\partial t}\right) + \dfrac{\rho_0 \vec{g}}{\tilde{C}_p^2}\dfrac{\partial \vec{V}}{\partial t} = -\vec{V}_0 \left(\dfrac{1}{\tilde{C}_p^2} + \dfrac{1}{\tilde{C}^2}\right)\nabla p \end{cases} \tag{3.1.7}$$

Let's find the expression $\nabla[\rho_0(\partial \vec{V}/\partial t)]$ from the first equation of system (3.1.7) and substitute it in the second one, we will obtain:

$$\frac{1}{\tilde{C}^2}\frac{\partial^2 p}{\partial t^2} + \frac{\vec{g}}{\tilde{C}^2}\left(\nabla p - \frac{2\nabla \tilde{C}}{\tilde{C}}p\right) - \Delta p + \frac{\rho_0 \vec{g}}{\tilde{C}_p^2}\frac{\partial \vec{V}}{\partial t} = \\ \nabla[\rho_0(\vec{V}_0 \nabla)\vec{V}] - \vec{V}_0\left(\frac{1}{\tilde{C}_p^2} + \frac{1}{\tilde{C}^2}\right)\frac{\partial}{\partial t}\nabla p \tag{3.1.8}$$

From the first equation of the system (3.1.1) we find that

$$\rho_0 \frac{\partial \vec{V}}{\partial t} = -\nabla p + \frac{\vec{g}}{\tilde{C}^2}p - \rho_0(\vec{V}_0 \nabla)\vec{V} \tag{3.1.9}$$

Substituting (3.1.9) in (3.1.8) we will get:

$$\frac{1}{\tilde{C}^2}\frac{\partial^2 p}{\partial t^2} + \vec{g}\left(\frac{1}{\tilde{C}^2} - \frac{1}{\tilde{C}_p^2}\right)\nabla p - \frac{\vec{g}}{\tilde{C}^2}\left(\frac{2\nabla \tilde{C}}{\tilde{C}} - \frac{\vec{g}}{\tilde{C}_p^2}\right)p - \Delta p =$$
$$\frac{\rho_0 \vec{g}}{\tilde{C}_p^2}(\vec{V}_0 \nabla)\vec{V} + \nabla[\rho_0(\vec{V}_0 \nabla)\vec{V}] - \vec{V}_0\left(\frac{1}{\tilde{C}_p^2} + \frac{1}{\tilde{C}^2}\right)\frac{\partial}{\partial t}\nabla p$$
(3.1.10)

If we take into consideration that

$$1/\tilde{C}^2 - 1/\tilde{C}_p^2 = 1/\tilde{C}_s^2 \qquad (3.1.11)$$

and ignore the third term in the left part of Equation (3.1.10) due to its obvious smallness, we will finally get

$$\frac{1}{\tilde{C}^2}\frac{\partial^2 p}{\partial t^2} + \frac{\vec{g}}{\tilde{C}_s^2}\nabla p - \Delta p = \frac{\rho_0 \vec{g}}{\tilde{C}_p^2}(\vec{V}_0 \nabla)\vec{V} + \nabla[\rho_0(\vec{V}_0 \nabla)\vec{V}] - \vec{V}_0\left(\frac{1}{\tilde{C}_p^2} + \frac{1}{\tilde{C}^2}\right)\frac{\partial}{\partial t}\nabla p \qquad (3.1.12)$$

The Equation (3.1.12) may be called the generalized equation of gravitational waves in ideal (nondissipative) and adiabatic media. In particular case, under corresponding values of coefficients, the equation of sound wave in the Earth atmosphere can be obtained from this equation. The first attempts of correct derivation of the equation of gravitational wave were done in the works [16] and [17], however, the equation of continuity of mass was used in the form of (2.2.3) and, consequently, their results differ from the results of the equation (3.1.12).

§2. Sound Wave as a Particular Case of the Gravitational Waves

For stationary medium $(V_0 = 0)$ equation (3.1.12) takes the form

$$\frac{1}{C^2}\frac{\partial^2 p}{\partial t^2} + \frac{\vec{g}}{C_s^2}\nabla p - \Delta p = 0 \qquad (3.2.1)$$

Let's consider the equation (3.2.1) for the waves propagating in horizontal $(k^2 = k_x^2 + k_y^2)$ and vertical $(k^2 = k_z^2)$ directions. In the first case p is represented in the form

$$p(x, y, t) = p_a \exp i(k_x x + k_y y - \omega t) \qquad (3.2.2)$$

after which $\vec{g}\nabla p = 0$ and the equation (3.2.1) transfers into equation of plane wave

$$\Delta p - \frac{1}{C^2}\frac{\partial^2 p}{\partial t^2} = 0 \qquad (3.2.3)$$

Substituting (3.2.2) into (3.2.3), we obtain

$$C^2(z,T) = \frac{\omega^2}{k^2} \qquad (3.2.4)$$

We see that the quantity $C(z,T)$, of which is determined by the formula (1.3.2), has the meaning of the phase velocity of the sound wave and therefore the equation (3.2.3) is the equation of the sound wave. Thus, the influence of the gravitational field on a sound wave propagating in the horizontal direction manifested in the fact that the speed of sound decreases with increasing altitude.

For waves propagating in vertical direction (along the axis Z) the equation (3.2.1) has to be solved by method of geometrical optics, since in this direction the atmosphere is non-homogeneous as a result of influence of gravitational field. Following this method [18] pressure perturbation will be presented in a form

$$p = p_a(z)\exp[i(-\omega t + \psi(z))] \qquad (3.2.5)$$

where $\psi(z)$ is a certain dimensionless function which satisfies the equation

$$\frac{d\psi(z)}{dz} = k_z(z) \qquad (3.2.6)$$

Then from equation (3.2.1) we will get

$$i\psi'' - (\psi')^2 + ig\frac{1}{C_s^2}\psi' + \frac{\omega^2}{C^2} = 0 \qquad (3.2.7)$$

The mark herein denotes derivative with respect to z. Using the expression for the upper boundary of the troposphere z_0, which is determined by formula (1.3.3), it will be easy to demonstrate that

$$C_p^2 = C_s^2 \frac{z_0^2}{z^2} \qquad (3.2.8)$$

$$C^2 = C_s^2 \frac{1}{1+z^2/z_0^2} \qquad (3.2.9)$$

Consequently, the equation (3.2.7) may be written as

$$i\psi'' - (\psi')^2 + i\frac{g}{C_s^2}\psi' + \frac{\omega^2}{C_s^2}\left(1 + \frac{z^2}{z_0^2}\right) = 0 \qquad (3.2.10)$$

Let's expand $\psi(z)$ along small parameter λ/z_0, where λ is the wave length and z_0 is the size of non-homogeneity of medium

$$\psi = \psi_0 + \psi_1 + \psi_2 + \ldots \ldots \quad (3.2.11)$$

and solve the equation (3.2.10) in null approximation for which let's suppose that $\psi = \psi_0$ and ignore the second derivative and thereafter we will obtain

$$\left(\psi_0'\right)^2 - i\frac{g}{C_s^2}\psi_0' - \frac{\omega^2}{C_s^2}\left(1 + \frac{z^2}{z_0^2}\right) = 0 \quad (3.2.12)$$

from which for ψ_0' we'll find

$$\psi_0' = \frac{1}{2C_s}\left[\frac{ig}{C_s} \pm 2\omega\left(1 + \frac{z^2}{z_0^2} - \frac{g^2}{4\omega^2 C_s^2}\right)^{1/2}\right] \quad (3.2.13)$$

From (3.2.13) it is seen that for propagation of sound along axis Z, the expression under the radical must be nonnegative, i.e. it is necessary to fulfill the condition

$$\omega \geq \left(\frac{g^2}{4C_s^2\left(1 + z^2/z_0^2\right)}\right)^{1/2} \quad (3.2.14)$$

The function $1/\left(1 + z^2/z_0^2\right)$ is close to one in the whole interval of change of z and therefore from (3.2.14) it follows that

$$\omega \geq \frac{g}{2C_s} \approx 1.85 \times 10^{-2}\, Hz \quad (3.2.15)$$

As we see sound propagation condition is fulfilled with big reserve and for sound frequencies ($\omega > 10\,Hz$), and after neglecting the terms above the first order of smallness (which is equivalent to neglecting of the second summand in equation (3.2.12)), the equation (3.2.13) will result in

$$\psi_0' = k_{0z}(z) = \pm\frac{\omega}{C_s}\left(1 + \frac{z^2}{z_0^2}\right)^{1/2} \quad (3.3.16)$$

Thus, sound wave spreading in vertical direction is also flat.

Let's now find $\psi_1'(z)$ and with this purpose let's insert $\psi = \psi_0 + \psi_1$ in (3.2.10). Having ignored the small quantity of the second order and taken into consideration the equation (3.2.12), we will obtain

$$\psi_1'(z) = \frac{i}{2}\frac{d}{dz}\ln\psi_0'(z) \tag{3.2.17}$$

From (3.2.17) it is obvious that in (3.2.16) the sign "+" should be taken. Suppose the sound source is in point $z = z_1$ $(0 \le z_1 \le z_0)$ and assume that in the stated interval temperature weakly depends on altitude, and therefore $C_s = const$. Then from (3.2.16) and (3.2.17) for $\psi_0(z)$ and $\psi_1(z)$ respectively, we will obtain

$$\psi_0(z) = \frac{\omega}{C_s}\int_{z_1}^{z}\left(1 + \frac{z'^2}{z_0^2}\right)^{1/2} dz' =$$

$$\frac{z_0\omega}{2C_s}\left\{\left(1+\frac{z^2}{z_0^2}\right)^{1/2}\frac{z}{z_0} - \left(1+\frac{z_1^2}{z_0^2}\right)^{1/2}\frac{z_1}{z_0} + \ln\frac{(1+z^2/z_0^2)^{1/2} + z/z_0}{(1+z_1^2/z_0^2)^{1/2} + z_1/z_0}\right\} \tag{3.2.18}$$

$$\psi_1(z) = \frac{i}{2}\int_{z_1}^{z} d\ln\psi_0'(z) = \frac{i}{2}\ln\frac{(1+z^2/z_0^2)^{1/2}}{(1+z_1^2/z_0^2)^{1/2}} \tag{3.2.19}$$

From (3.2.5), (3.2.18) and (3.2.19) for pressure perturbation in first approximation we will have

$$p(z;t) = \tilde{p}(z)\exp i\omega\left\{\frac{z_0}{2C_s}\left[\left(1+\frac{z^2}{z_0^2}\right)^{1/2}\frac{z}{z_0} - \left(1+\frac{z_1^2}{z_0^2}\right)^{1/2}\frac{z_1}{z_0}\right] - t\right\} \tag{3.2.20}$$

where

$$\tilde{p}(z) = P_0\left(\frac{1+z_1^2/z_0^2}{1+z^2/z_0^2}\right)^{1/4} \tag{3.2.21}$$

From (3.2.21) we see that when the sound is propagated from the bottom to the top the wave amplitude drops while when it is propagated from the top to the bottom it increases. Thus, for instance, if the sound source is at sea level $(z_1 = 0)$, then by reaching altitude $z = z_0$, wave amplitude $\tilde{p}(z_0) = 0{,}84 P_0$.

Let's now find $k_z(z)$, which by determination is $k_z = \psi_0' + \psi_1'$ and using expressions (3.2.16) and (3.2.17) we will find

$$k_z(z) = \frac{\omega}{C_s}\left(1+\frac{z^2}{z_0^2}\right)^{1/2} + \frac{i}{2}\frac{z}{z_0^2(1+z^2/z_0^2)} \tag{3.2.22}$$

Module of imaginary component k_z achieves largest extremum in the point $z = z_0$ and equals to $(k_z)_{max} = 1/4z_0 \approx 10^{-4}$ m^{-1} which allows to ignore it as compared to the real part and finally we will get

$$k_z(z) = \frac{\omega}{C_s}\left(1+\frac{z^2}{z_0^2}\right)^{1/2} = \frac{\omega}{C(z)} \qquad (3.2.23)$$

where $C(z) = C_s\left(1+z^2/z_0^2\right)^{-1/2}$ and $C_s = const$.

As we see, the sound wave equation in any direction has the form of the equation of plane wave and the wave number in vertical direction is analytically expressed by the formula similar to the horizontal direction. However, there is difference between them. In the second case the wave number at the stated altitude has a constant value while in the first case it increases when sound is propagated from the bottom to the top and decreases in case of top-to-bottom propagation. Such behavior of $k_z(z)$ serves as a reason for the phenomenon known as "refraction of sound" the essence of which is that while the sound wave propagates in the horizontal direction, the wave vector changes its direction to the side of decrease of the sound speed. In the current theory this phenomenon is explained by gradient along the axis Z of adiabatic speed of sound C_s which is conditioned by its dependence on temperature T, which in its turn depends on coordinate z according to the aforementioned law. Adhering to our theory, gravitational field of the Earth also contributes to sound refraction which is expressed by emergence of multiplier $G(z) = \left(1+z^2/z_0^2\right)^{1/2}$ in (3.2.23). For assessment of this contribution it is necessary to calculate $dk_z(z)/dz$. Error made by averaging of $C_s(T)$ is small and, therefore, in (3.3.23) we can admit $C_s = \sqrt{\gamma RT/M}$, where $T = -6{,}52 \cdot 10^{-3} z + 288{,}15$ and then we get:

$$\frac{dk_z(z)}{dz} = \omega \frac{zT + 3.26 \times 10^{-3} G^2(z) z_0^2}{z_0^2 T G(z) C(z)} \qquad (3.2.24)$$

Effect of gravitational filed takes into consideration the first term in numerator (3.2.24), influence of which becomes determining when

$$z \geq \frac{3{,}26 \times 10^{-3} G^2(z) z_0^2}{T} \approx 2{,}3 \times 10^3 \, m \qquad (3.2.25)$$

Here we assume that $G^2(z) = 4/3$ and $T = 252{,}4^\circ K$ are the average values of these quantities. Apparently this is the altitude up to which the sound can be considered adiabatic and above which influence of gravitational filed cannot be ignored.

§ 3. Peculiarities of Sound Wave in Moving Medium

As it is known, the law of conservation of mass being one of the fundamental laws of physics was formulated for homogeneous medium. However, it is still considered as universal law and up to now it has been applied to inhomogeneous medium like the Earth atmosphere [3]-[6], [8] or the ocean [7], [19;20]. Above, it was demonstrated that after having taken into consideration the effect of the gravitational field of the Earth this equation needs to be perfected. It resulted in generalizing of great number of known gas or hydrodynamic relations, as well as in new understanding of the most important properties of medium like compressibility and incompressibility. In this regard, it was a logical necessity to verify applicability of fundamental laws of physics such as Doppler's effect and principle of relativity of motion to inhomogeneous medium. For that purpose, we consider the equation (3.1.12). Let's assume that $\vec{V}_0 = V_0 \vec{e}_x$ where $V_0 > 0$, $\vec{e}_x$ is the unit vector directed along the axis X and $\rho_0 = \rho_0(z)$. Then it is easy to check that

$$\frac{\rho_0 \vec{g}}{\tilde{C}_p^2}(\vec{V}_0 \nabla)\vec{V} = -\frac{\rho_0 V_0 g}{\tilde{C}_p^2}\frac{\partial V_z}{\partial x} \tag{3.3.1}$$

and

$$\nabla[\rho_0(\vec{V}_0 \nabla)\vec{V}] = V_0\left[\rho_0 \frac{\partial^2 V_x}{\partial x^2} + \frac{\partial}{\partial z}\left(\rho_0 \frac{\partial V_z}{\partial x}\right)\right] \tag{3.3.2}$$

Substituting both (3.3.1) and (3.3.2) into (3.1.12), we will obtain the following expression:

$$\frac{1}{\tilde{C}^2}\frac{\partial^2 p}{\partial t^2} - \frac{g}{\tilde{C}_s^2}\frac{\partial p}{\partial z} - \Delta p =$$
$$V_0\left[\rho_0 \frac{\partial^2 V_x}{\partial x^2} + \frac{\partial}{\partial z}\left(\rho_0 \frac{\partial V_z}{\partial x}\right) - \frac{\rho_0 g}{\tilde{C}_p^2}\frac{\partial V_z}{\partial x} - \left(\frac{1}{\tilde{C}_p^2} + \frac{1}{\tilde{C}^2}\right)\frac{\partial^2 p}{\partial x \partial t}\right] \tag{3.3.3}$$

Let's find all variables in the form of plane wave $f(\vec{r},t) = f_a(z)\exp[i(kx - \omega t)]$ where $f_a(z)$ is the amplitude. Then, we will obtain from (3.3.3) the following

$$\frac{\omega^2}{\tilde{C}^2}p_a(z) + \frac{g}{\tilde{C}_s^2}\frac{dp_a(z)}{dz} + \frac{d^2 p_a(z)}{dz^2} - k^2 p_a(z) =$$
$$V_0\left[k^2 \rho_0 V_{ax}(z) - ik\frac{d}{dz}(\rho_0 V_{az}(z)) + ik\frac{\rho_0 g}{\tilde{C}_p^2}V_{az}(z) + \omega k\left(\frac{1}{\tilde{C}_p^2} + \frac{1}{\tilde{C}^2}\right)p_a(z)\right] \tag{3.3.4}$$

For x and z components of the first equation of system (3.1.1), we will easily find

$$\rho_0 V_{ax}(z) = \frac{k p_a(z)}{\omega - kV_0} \tag{3.3.5}$$

$$\rho_0 V_{az}(z) = -\frac{i}{\omega - kV_0}\left(\frac{dp_a(z)}{dz} + \frac{g}{\tilde{C}^2}p_a(z)\right) \tag{3.3.6}$$

Differentiating equation (3.3.6) with respect to z, we will get

$$\frac{d}{dz}(\rho_0 V_{az}(z)) = -\frac{i}{\omega - kV_0}\left(\frac{d^2 p_a(z)}{dz^2} + \frac{g}{\tilde{C}^2}\frac{dp_a(z)}{dz} - \frac{2g}{\tilde{C}^3}\frac{d\tilde{C}}{dz}p_a(z)\right) \tag{3.3.7}$$

Having substituted (3.3.5), (3.3.6) and (3.3.7) in (3.3.4) and having ignored the same summand as in (3.1.10), we will obtain equation for amplitude of linear gravitational wave in moving inhomogeneous medium in the form of

$$\frac{d^2 p_a(z)}{dz^2} + \frac{g}{\tilde{C}_s^2} \frac{dp_a(z)}{dz} + \left[\frac{(\omega - kV_0)^2}{\tilde{C}^2} - \frac{kV_0(\omega - kV_0)}{\tilde{C}_p^2} - k^2 \right] p_a(z) = 0 \qquad (3.3.8)$$

Let's consider sound wave when $p_a(z) = const$. In this case the dispersive equation (3.3.8) takes the form of

$$\frac{(\omega - kV_0)^2}{\tilde{C}^2} - \frac{kV_0(\omega - kV_0)}{\tilde{C}_p^2} - k^2 = 0 \qquad (3.3.9)$$

When $V_0 = 0$ equation (3.3.9) results in

$$\omega^2 = k^2 \tilde{C}^2 = k^2 C^2 \qquad (3.3.10)$$

and thus homogeneity (compressibility, $(C = C_s)$) or heterogeneity (incompressibility, $(C = C_p)$) of the medium at rest have no influence on propagation of sound wave. If medium is immovable and the source of sound is moving towards the receiver with speed of U_0, then, as a result of change of length of the wave $k \to \tilde{k} = \omega/(C - U_0)$, the wave frequency which perceived by the receiver will equal to

$$\tilde{\omega} = \tilde{k} C = \frac{\omega}{1 - U_0/C} \qquad (3.3.11)$$

i.e. frequency shift in any medium occurs according to Doppler's law [21]. Substitution of U_0 by $-U_0$ in (3.3.11) corresponds to motion of the source from the receiver. Consequently, Doppler's law and the principle of motion relativity for sound waves are invariant in respect of properties of immovable medium.

When medium is moving (fixed source and receiver moving with medium) the picture essentially changes. From (3.3.9) it is clear that in moving inhomogeneous medium, shift of frequency occurs according to the law different from Doppler's law.

$$(\omega - kV_0) \left[\omega - kV_0 \left(1 + \frac{\tilde{C}^2}{\tilde{C}_p^2} \right) \right] = k^2 \tilde{C}^2 \qquad (3.3.12)$$

Equation (3.3.12) is fair for adiabatic medium representing the troposphere, where speed of the wind is $V_0 \leq 30 \, m/sec$ and since $V_0 \ll C_s \leq C_p$ the following inequality is fair

$$\frac{\tilde{C}}{\tilde{C}_p} \approx \frac{C}{C_p} \approx \frac{1}{\sqrt{1+C_p^2/C_s^2}} \leq 1 \qquad (3.3.13)$$

Equality sign in (3.3.13) is fair only for strongly inhomogeneous medium, when $C_s \gg C_p$ and $C \approx C_p$. When $C_s \ll C_p$ and $C \approx C_s$ relation $\tilde{C}^2/\tilde{C}_p^2$ in (3.3.12) can be ignored and then we will have

$$\tilde{\omega}^2 = (\omega - kV_0)^2 = \omega^2(1 - V_0/C_s)^2 = k^2(C_s - V_0)^2 = k^2\tilde{C}_s^2 \qquad (3.3.14)$$

As we can see, in weakly inhomogeneous medium, shift of frequency occurs according to Doppler's law when the receiver moves towards the source and the wave length does not change. Substitution in (3.3.14) $-V_0$ by V_0 which is equivalent to the case when the receiver moves from the source, results in the following

$$\tilde{\omega}^2 = (\omega + kV_0)^2 = \omega^2(1 + V_0/C_s)^2 = k^2(C_s + V_0)^2 = k^2\tilde{C}_s^2 \qquad (3.3.15)$$

Thus, for weakly inhomogeneous moving medium ($C_s \ll C_p$) Doppler's law and the principle of relativity of motion are fair. This case is realized in the Earth atmosphere, for instance, up to the altitude of approximately $z = 3$ km, where $C_s \approx 328,6$ m/sec and $C_p \approx 1355,3$ m/sec and consequently $C^2/C_p^2 \approx C_s^2/C_p^2 \approx 0,06$. At the upper boundary of troposphere, where $C_p = C_s$ [9;12] and $C^2 = 0,5C_p^2$, the equation (3.3.12) results in

$$(\omega - kV_0)(\omega - 0,5kV_0) = k^2(C - V_0)^2, \qquad (3.3.16)$$

while for strongly inhomogeneous medium $(C_s \gg C_p)$ we have

$$(\omega - kV_0)(\omega - 2kV_0) = k^2(C - V_0)^2, \qquad (3.3.17)$$

We see that equality of (3.3.16) and (3.3.17) describes sound wave only when $V_0 = 0$. It means that in moving inhomogeneous medium, where isobaric speed of sound is at least to some extent significant, propagation of sound is impossible. This is explained by the mechanism of propagation of isobaric sound in which perturbation of density is spread as a result of change of the volume of fixed mass of liquid or gas. In the moving inhomogeneous medium, the concept of constant mass in any volume loses its meaning. This is evidenced by the formula (2.3.2) for change of the mass of liquid (gas) in inhomogeneous medium. This result is obtained in the work [14].

Results:

Based on the new definition of the criteria of compressibility and incompressibility of medium

and generalized mass continuity equation in this chapter the following results is obtained :
- A new equation for linear gravity waves in a moving inhomogeneous medium has been received.
- It is shown that from this equation, in a special case, an equation of sound waves is received and by this it is proved that the sound wave in the Earth's atmosphere is a gravitational wave.
- It is discovered the influence of the gravitational field on a sound wave that is presented in the fact that during propagation of sound in a horizontal direction, its speed at a given height is constant and decreases with increasing height. During propagation in the vertical direction from the bottom to the top, speed of sound and the amplitude of perturbation of pressure are dropping, and during propagation from the top down- are growing.
- It is discovered that the influence of the gravitational field on the refraction of sound becomes significant for heights $z \geq 2,3$ km.
- It is shown that Doppler's law and the principle of relativity of motion are valid in relation to the sound waves for strongly inhomogeneous medium, if the medium is at rest.
- It is discovered that the propagation of sound in a moving strongly inhomogeneous medium is not possible

These results are published in works [22; 23].

CHAPTER IV

Disadvantages of theory of tangential discontinuity and how to overcome them

The Problems

Phenomenon of tangential discontinuity is widespread in nature. Tangential discontinuity is called a sharp change of the parameters of the two media at intersection of the surface separating these environments when the vectorial parameters are directed along the surface. Surfaces of tangential discontinuity are surfaces of water bodies (seas and oceans), surface of jets as usual so a conductive liquids, the surface of various layers of the magnetosphere and the surface of the magnetosphere as a whole, when the Solar wind flows around it and many others. If such a surface is perturbed in any small area, this perturbation will be distributed along it in the form of growing or decaying surface wave. In the first case tangential discontinuity called unstable , and in the second case- stable. Thus, the study of the stability or instability conditions of tangential discontinuity have of great practical importance in hydrodynamics and in magnetohydrodynamics.

As the analysis of the existing literature shows , these studies are carried out incorrectly, and usually lead to paradoxical, mutually exclusive results. This circumstance was first pointed in 1994 in the work [24], but for some reason it did not attract enough attention of the scientific community. In this chapter we will present some results mentioned in this work and as well as the new ones obtained in recent years.

§ 1. Incorrectness of the modern theory of hydrodynamic tangential discontinuity

The simplest task of hydrodynamic tangential discontinuity in the linear approximation was solved by Helmholts in 1868 ([3], §29). He considered tangential discontinuity on a flat surface $z = 0$ between of two incompressible fluid without taking into account the forces of surface tension and gravity, whose densities respectively are ρ_{01} $(z>0)$ and ρ_{02} $(z<0)$. The upper fluid moves relatively to the low one at the speed V_0 along the axis X $(\vec{V}_0 = V_0 \vec{e}_x)$. In the task the incompressibility is understood in conventional sense, when $\nabla \vec{V} = 0$ and $\tilde{C} = \infty$. As it follows from abovesaid, assumption of incompressibility of liquids ($\nabla \vec{V} = 0$) is already erroneous.

The equation for the perturbed pressure value can be obtained from the generalized linear equation of gravitational waves (3.1.12), and therefore let us write it again:

$$\frac{1}{\tilde{C}^2}\frac{\partial^2 p}{\partial t^2} + \frac{\vec{g}}{\tilde{C}_s^2}\nabla p - \Delta p = \frac{\rho_0 \vec{g}}{\tilde{C}_p^2}(\vec{V}_0 \nabla)\vec{V} + \nabla[\rho_0(\vec{V}_0 \nabla)\vec{V}] - \vec{V}_0\left(\frac{1}{\tilde{C}_p^2} + \frac{1}{\tilde{C}^2}\right)\frac{\partial}{\partial t}\nabla p \quad (3.1.12)$$

According to (3.1.6) from the condition $\tilde{C} = \infty$ it follows that $\tilde{C}_s = \tilde{C}_p = \infty$ and from (3.1.12), we obtain:

$$\Delta p = -\nabla[\rho_0(\vec{V}_0 \nabla)\vec{V}] \quad (4.1.1)$$

As we can see, if the incompressibility is understood in the sense in which it is treated in modern theory, the summand, containing the gravitational acceleration, disappears from the equation (3.1.12), and it means that the gravitational wave, as any other one, in incompressible fluid does not exist. Moreover, taking into account that $\rho_0 = const$ and $\nabla \vec{V} = 0$ from the equation (4.1.1), we obtain:

$$\Delta p = -V_0\left[\frac{\partial \vec{V}}{\partial x}\nabla \rho_0 + \rho_0 \frac{\partial}{\partial x}div\vec{V}\right] = 0 \quad (4.1.2)$$

Thus, in the liquid where the density is constant, pressure perturbation is determined from Laplace's equation $\Delta p = 0$. Perturbation of pressure p is represented as

$$p(x,z,t) = p_a(z)\exp[i(kx - \omega t)] \quad (4.1.3)$$

Requiring damping of amplitude of surface wave at removing from the surface of discontinuity ($p_a(z) \to 0$ when $z \to \pm\infty$) from (4.1.2) and (4.1.3) we obtain:

$$\begin{cases} p_1 = const \exp(-kz)\exp[i(kx - \omega t)](z > 0) \\ p_2 = const \exp(kz)\exp[i(kx - \omega t)](z < 0) \end{cases} \quad (4.1.4)$$

Let $\xi = \xi(x,t)$ be a displacement of the points of discontinuity surface along the axis Z under perturbation. Then it is obvious that

$$V_z = \frac{d\xi}{dt} = \frac{\partial \xi}{\partial t} + V_0 \frac{\partial \xi}{\partial x} \tag{4.1.5}$$

If ξ is represented in the form

$$\xi(x,t) = \xi_a \exp[i(kx - \omega t)] \tag{4.1.6}$$

then from (4.1.5), we find

$$V_z = i\xi(kV_0 - \omega) \tag{4.1.7}$$

On the basis of (4.1.6) and (4.1.7) from the z component of Euler's equation (1.1.1) when $g = 0$ we obtain

$$\frac{\partial p}{\partial z} = \rho_0 (kV_0 - \omega)^2 \xi \tag{4.1.8}$$

Substituting in (4.1.8) values p_1 and p_2 from (4.1.4) and taking into account that $V_0 = 0$ in the range of $z < 0$, we finally get:

$$p_1 = -\xi \frac{\rho_{01}(kV_0 - \omega)^2}{k} \qquad (z > 0) \tag{4.1.9}$$

$$p_2 = \xi \frac{\rho_{02}\omega^2}{k} \qquad (z < 0) \tag{4.1.10}$$

From the condition of equality p_1 and p_2 on the surface of discontinuity we get the dispersion equation for gravitational waves in the form of:

$$\rho_{01}(kV_0 - \omega)^2 = -\rho_{02}\omega^2 \tag{4.1.11}$$

solution of which is

$$\omega = kV_0 \frac{\rho_{01} \pm i\sqrt{\rho_{01}\rho_{02}}}{\rho_{01} + \rho_{02}} \tag{4.1.12}$$

We can see that the roots of the equation (4.1.11) are the two conjugative complex numbers, for one of which $\text{Im}\{\omega\} > 0$ and as follows from (4.1.3) for such perturbations $p_{1,2} \to \infty$ when $t \to \infty$, i.e. such tangential discontinuity is absolutely unstable.

In 1944 Landau solved Helmholz's problem for the compressible fluids ([2], §84). When $g = 0$ in (3.1.12), it is necessary to assume that $\tilde{C}_p = \infty$ and in that case we have:

$$\frac{1}{\widetilde{C}^2}\frac{\partial^2 p}{\partial t^2} - \Delta p = \rho_0 V_0 \frac{\partial}{\partial x}(\nabla \vec{V}) - \frac{\vec{V_0}}{\widetilde{C}^2}\frac{\partial}{\partial t}\nabla p \qquad (4.1.13)$$

From the second equation (3.1.3) according to the above mentioned assumptions, we obtain:

$$\rho_0(\nabla \vec{V}) = -\frac{1}{\widetilde{C}^2}\left(\frac{\partial p}{\partial t} + \vec{V_0}\nabla p\right) = -\frac{1}{\widetilde{C}^2}\left(\frac{\partial p}{\partial t} + V_0\frac{\partial p}{\partial x}\right). \qquad (4.1.14)$$

Substituting (4.1.14) into (4.1.13), we get the wave equation for the value of perturbed pressure in the form of

$$\Delta p = \frac{1}{\widetilde{C}^2}\left[\frac{\partial^2 p}{\partial t^2} + 2V_0\frac{\partial^2 p}{\partial x \partial t} + V_0^2\frac{\partial^2 p}{\partial x^2}\right] \qquad (4.1.15)$$

Landau looked for a solution of the equation (4.1.15) in the ranges of $z > 0$ and $z < 0$ correspondingly in the form of

$$\begin{cases} p_1 = const \exp(-i\omega t + ikx - i\chi_1 z)(z > 0) \\ p_2 = const \exp(-i\omega t + ikx + i\chi_2 z)(z < 0) \end{cases} \qquad (4.1.16)$$

where $\chi_1 = \alpha_1 + i\beta_1$ and $\chi_2 = \alpha_2 + i\beta_2$ are the complex numbers. If we compare the expressions (4.1.16) with the expressions (4.1.4), we will see that in this case the amplitudes of the perturbation of pressures in the ranges of $z > 0$ and $z < 0$ are

$$p_{a1} = const \exp(-i\chi_1 z) = const \exp(-i\alpha_1 z)\exp(\beta_1 z) \quad (z > 0) \qquad (4.1.17)$$

$$p_{a2} = const \exp(i\chi_2 z) = const \exp(i\alpha_1 z)\exp(-\beta_2 z) \quad (z < 0) \qquad (4.1.18)$$

Presentation of perturbations amplitudes of pressures in the form of (4.1.17) and (4.1.18) is due to the requirement of perturbation attenuation at removing from the surface of discontinuity, i.e. $p_{a1}(z) \to 0$ when $z \to \infty$ and $p_{a2}(z) \to 0$ when $z \to -\infty$ (surface waves). Based on these requirements the authors of this monograph rightly believe that $\beta_2 < 0$ but they say nothing about the sign β_1 which should also be negative, i.d. $\beta_1 < 0$ Substituting (4.1.16) into (4.1.15) and considering the fact that $V_0 = 0$ in the range $z < 0$ for χ_1^2 and χ_2^2, we obtain:

$$\chi_1^2 = \frac{(kV_0 - \omega)^2}{\widetilde{C}_1^2} - k^2 \qquad (4.1.19)$$

$$\chi_2^2 = \frac{\omega^2}{C_2^2} - k^2 \qquad (4.1.20)$$

Values of components of perturbed velocity are given by formulas similar to (4.1.16), after which from z component of the Euler equation (1.1.1) we obtain the following:

$$V_{z1} = \frac{\chi_1 p_1}{\rho_1(kV_0 - \omega)} \quad (z > 0) \qquad (4.1.21)$$

$$V_{z2} = \frac{\chi_2 p_2}{\rho_2 \omega} \quad (z < 0) \quad ? \qquad (4.1.22)$$

After this, by calculations similar to Helmholtz task, for perturbed pressures author receive

$$p_1 = -\frac{\rho_1(kV_0 - \omega)^2 \xi}{i\chi_1} \qquad (4.1.23)$$

$$p_2 = \frac{\rho_2 \omega^2 \xi}{i\chi_2} \qquad (4.1.24)$$

The equality of these two variables on the surface of discontinuity gives dispersion equation in the form of:

$$\frac{\chi_1}{\rho_1(kV_0 - \omega)^2} = -\frac{\chi_2}{\rho_2 \omega^2} \qquad (4.1.25)$$

Substituting in (4.1.25) the values of χ_1 and χ_2 from formulas (4.1.19) and (4.1.20), subject to $\rho_1 = \rho_2 = \rho$ and $\tilde{C}_1 = C_1 = C_2 = C$, we obtain a transcendental equation for ω

$$\frac{(kV_0 - \omega)^2}{[(kV_0 - \omega)^2 / C^2 - k^2]^{1/2}} = -\frac{\omega^2}{[\omega^2 / C^2 - k^2]^{1/2}} \qquad (4.1.26)$$

After squaring of the equation (4.1.26), Landau received the equation

$$\left[\frac{1}{\omega^2} - \frac{1}{(kV_0 - \omega)^2}\right]\left[\frac{1}{C^2 k^2} - \frac{1}{\omega^2} - \frac{1}{(kV_0 - \omega)^2}\right] = 0 \qquad (4.1.27)$$

solutions of which is

$$\omega = \frac{kV_0}{2} \qquad (4.1.28)$$

$$\omega = \frac{k}{2}\left[V_0 \pm \left(V_0^2 + 4C^2 \pm 4C\sqrt{C^2 + V_0^2}\right)^{1/2}\right] \qquad (4.1.29)$$

For the reasons stated above, tangential discontinuity in a compressible liquid is stable, if all roots of the equation (4.1.26) are real numbers, and as follows from (4.1.29) it is possible if the following condition is satisfied:

$$V_0^2 + 4C^2 - 4C\sqrt{C^2 + V_0^2} \geq 0 \qquad (4.1.30)$$

or finally

$$V_0 \geq \sqrt{8}C \qquad (4.1.31)$$

This condition is diametrically contrary to the stability condition for incompressible fluids, for it says that in a compressible fluid at a collinear vectors $\vec{k}$ and $\vec{V}_0$, tangential discontinuity is stable if the relative flow velocity exceeds the speed of sound $\sqrt{8}$ or more times, while from the formula (4.1.12) follow that in an incompressible fluid, an arbitrarily small value of the relative speed of flow leads to instability of the tangential discontinuity. To avoid this contradiction, in the footnote section the authors write - "The value of (4.1.31) was obtained by Landau (1944). In this problem, the necessity of considering noncollinearity of vectors $\vec{V}_0$ and $\vec{k}$ was indicated by Syrovatski (1954)." After taking into account Syrovatskii's indication [25], everywhere, including in the equation (4.1.12), instead kV_0 we have $kV_0\cos\varphi$, where φ is the angle between vectors $\vec{k}$ and $\vec{V}_0$ and condition (4.1.31) takes the form

$$V_0\cos\varphi \geq V_{cr} = \sqrt{8}C \qquad (4.1.32)$$

According to the authors, after the adoption of the conditions (4.1.32) there is no contradiction, as evidenced by the following arguments: "Thus, when $V_0\cos\varphi < V_{cr}$ the dispersion equation has a pair of complex-conjugate roots, for one of which will be $\text{Im}\{\omega\} > 0$ and relevant perturbations lead to instability. If $V_0 < V_{cr}$ such perturbations are perturbations with any angle, and when $V_0 > V_{cr}$, unstable are only perturbations for which $\cos\varphi < V_{cr}/V_0$. As a result, the tangential discontinuity is always unstable. Let's note that the fact of instability (if we are not interested for which perturbations) is obvious already from the instability, in the case of an incompressible fluid, coupled with the fact that the dispersion equation includes velocity V_0 only in combination of $V_0\cos\varphi$ and for any speed V_0, there are always angles φ, for which $\cos\varphi < V_{cr}/V_0$, so that in relation to such perturbations the medium behaves as an incompressible. "

Introducing the dimensionless quantities: $\phi = U_p/C$, $M = V_0k_x/kC = V_0 \sos\varphi/C$, when $U_p = \omega/k$ is the phase velocity of the wave of perturbation and $k = \sqrt{k_x^2 + k_y^2}$, equation (4.1.26) can be reduced to an equation of the form:

$$\phi^2[1-(\phi = M)^2]^{1/2} = -(\phi - M)^2(1-\phi^2)^{1/2} \qquad (4.1.33)$$

This equation, in 1968 was solved by Gerwin with numerical method [26]. The amplitudes of pressure perturbations for moving fluid and fluid at rest are respectively, proportional to the exponents:

$$\begin{cases} p_1 \approx \exp\left\{-\left[1-(\phi-M)^2\right]^{1/2}kz\right\} \\ p_2 \approx \exp\left[(1-\phi^2)^{1/2}kz\right] \end{cases} \qquad (4.1.34)$$

The necessary condition for stability is the existence of real roots for equation (4.1.33) which is impossible if the expressions under the radicals are positive due to the sign "minus" in the right side of this equation. Gerwin believes that such roots may exist only for the case when the following conditions are simultaneously fulfilled:

$$\begin{cases} 1-(\phi-M)^2 < 0 \\ 1-\phi^2 < 0 \end{cases} \qquad (4.1.35)$$

as well as when corresponding sign is chosen in front of the radicals in equation (4.1.33). Starting from this point of view, Gerwin numerically analyzes the equation

$$\phi^2\left[(\phi-M)^2-1\right]^{1/2} - (\phi-M)^2(\phi^2-1)^{1/2} = 0 \qquad (4.1.36)$$

and arrives at the conclusion that it has real roots if the effective Mach number satisfies inequality $M > 2^{3/2}$ or

$$V\cos\varphi > 2^{3/2}C \qquad (4.1.37)$$

which coincides with the condition of Landau-Syrovatsky (4.1.32).

§ 2. The correct solution to the task of Landau

As we see, authors of the monograph [3] bring forward the arguments which, in their opinion, eliminate contradictions which have arisen at the solution of the problem of hydrodynamic tangential discontinuity for incompressible and compressible liquids. These arguments are unacceptable for many reasons, one of which is shown above, but there is another, no less important. The condition $\cos\varphi < V_{cr}/V_0$ is satisfied for all V_0 if, $\varphi = \pi/2$ i.e. perturbations that propagate perpendicular to the flow in compressible fluids are the most instable, while in an incompressible fluid only such perturbations are stable, because for them $\omega = 0$.

Thus, the contradictions are not eliminated, because their reasons are not revealed. The reason is that the transformation by which equation (4.1.27) was obtained, is not identity, and thus, this equation does not correspond to the assigned task and determination of conditions of stability from it is unlawful. Analysis of the equation (4.1.25), which is the similar to the equation (4.1.26), shows that they can not have real roots. Indeed, as it can be seen from (4.1.19) and (4.1.20), if ω is the real number, then χ_1^2 и χ_2^2 are real too and consequently, χ_1 и χ_2 are positive real numbers. In such a case $\beta_1 = \beta_2 = 0$ and as it follows from (4.1.17), (4.1.18), the conditions of perturbation attenuation with distance from the surface of discontinuity are not performed. Thus, it is necessary to require the negativity of χ_1^2 and χ_2^2, i.e. $\chi_1 = i\beta_1$ and $\chi_2 = i\beta_2$, whereupon the equation (4.1.25) takes the form

$$\frac{\beta_1}{(kV_0-\omega)^2} = -\frac{\beta_2}{\omega^2} \qquad (4.2.1)$$

We noted above that the conditions of the task require fulfillment of inequalities $\beta_1 < 0$ and $\beta_2 < 0$ and as it is easy to see that, in this case, the equation (4.1.33) has no real roots, and hence the tangential discontinuity in a compressible fluid is absolutely unstable regardless of the direction of the disturbance propagation. Gerwin made the same mistake as Landau did, because the squaring equation (4.1.26) is the same mistake as choosing the sign "minus" in front of one of the radicals in the equation (4.1.36).

§ 3. Incorrectness of modern linear theory of surface capillary-gravity waves

Problems of tangential discontinuity at the interface between the two liquids which are considered above, were solved without taking into account the forces of surface tension and gravity, which does not correspond to reality. We have shown that such tangential discontinuities is absolutely unstable as in incompressible as well as in compressible fluids. Our results also imply that such classification of fluids is not correct, because one should speak of a tangential discontinuity in homogeneous and heterogeneous fluids. Since in these tasks the influence of the gravitational field of the Earth are not taken into account, they all belong to a homogeneous fluid, where in the first case the adiabatic speed of sound is lot more than speed of flow, and in the second case, they are of the same order.

Linear waves on the surface of a tangential discontinuity, which are generated and distributed under the influence of forces of surface tension and gravity, are called capillary-gravity surface waves. As it is known two limit cases are considered for such waves:

1. Shortwave disturbances when the wave length λ is much smaller than the fluid depth h ($kh \gg 1$, where $k = 2\pi/\lambda$ is wave number). In this case, the water is considered infinitely deep and the influence of surface tension of fluid is taken into account. Waves generated in such conditions on the water surface are called waves in deep water or capillary-gravity waves.

2. Longwave disturbances when the wave length λ is much greater than the fluid dept h ($kh \ll 1$). In this case the surface tension influence is ignored and generated waves are called waves on shallow water.

According to the existing theory, the spectrum of frequencies of capillary-gravity waves is calculated by the formula

$$\omega = \sqrt{k\left(g + \frac{\alpha k^2}{\rho}\right) th(kh)} \qquad (4.3.1)$$

where α is coefficient of surface tension of water, ρ is its density while g is acceleration of gravity. If the requirement is met

$$k \ll (\rho g / \alpha)^{1/2} \qquad (4.3.2)$$

capillarity effect can be neglected. Taking into consideration that for water $\alpha = 0{,}073$ N/m and

$\rho = 10^3 \text{kg/m}^3$, it is easy to calculate that the wave, length of which is $\lambda > 1,73$ cm, is purely gravitational, frequency of which in deep water $(th(kh) \approx 1)$ is

$$\omega = \sqrt{kg} \qquad (4.3.3)$$

In the second case, taking into consideration that $\alpha = 0$ and $th(kh) \approx kh$, we will have

$$\omega - k\sqrt{gh} \to U_p - \sqrt{gh} \qquad (4.3.4)$$

These formulae form the basis for large number of fundamental researches and are widely used in solution of applied problems. For example, phase speed of tsunami waves is calculated according to formula (4.3.4) [19;20].

Correlation of (4.3.1.), (4.3.3) and (4.3.4) follows from the Kelvin theory who fisrt solved the problem of surface capillary-gravity waves in 1871 ([3], see §62), i.e. generalized Helmholtz problem by taking into account the forces of surface tension and gravity. Assuming that the flow is potential, Kelvin introduced the velocity potential $\varphi(\vec{r},t)$ satisfying the Laplace equation

$$\Delta\varphi(\vec{r},t) = 0 \qquad (4.3.5)$$

and related to the flow velocity of incompressible fluid by equation

$$\vec{V} = \nabla\varphi \qquad (4.3.6)$$

The fluid pressure is defined by integration of the Euler's equation with application of (4.3.6), as well as condition of motion potentiality $rot\vec{V} = 0$ ([3], see §9):

$$p = -\rho\frac{\partial\varphi}{\partial t} - \rho gz - \rho\frac{V^2}{2} \qquad (4.3.7)$$

where z is the coordinate along the axis normal to the surface of discontinuity. Assuming that the upper fluid moves with respect to the lower fluid along axis X with velocity V_0, the velocity potentials satisfying equation (4.3.5) are written in the form

$$\varphi_1 = A_1 e^{-kz} \cos(kx - \omega t) + V_0 x, \qquad z > 0 \qquad (4.3.8)$$
$$\varphi_2 = A_2 e^{+kz} \cos(kx - \omega t), \qquad z < 0 \qquad (4.3.9)$$

The difference of signs in the exponents is due to the requirement of attenuation of disturbances while removing from discontinuity surface (surface waves). The constants A_1 and A_2 are considered as small quantities since they correspond to the disturbed values of the potentials. At the discontinuity surface the following conditions must be fulfilled:

$$(p_2 - p_1)\big|_{z=0} = -\alpha \frac{\partial^2 \xi}{\partial x^2} \qquad (4.3.10)$$

$$v_{z1}\big|_{z=0} = \frac{d\xi}{dt} = \frac{\partial \xi}{\partial t} + V_0 \frac{\partial \xi}{\partial x} \qquad (4.3.11)$$

$$v_{z2}\big|_{z=0} = \frac{\partial \xi}{\partial t} \qquad (4.3.12)$$

Condition (4.3.10) with consideration of (4.3.7) Kelvin writes as follows:

$$\rho_2 \frac{\partial \varphi_2}{\partial t} + \rho_2 g \xi - \alpha \frac{\partial^2 \xi}{\partial x^2} = \rho_1 \frac{\partial \varphi_1}{\partial t} + \rho_1 g \xi + \frac{\rho_1}{2}\left(V_1^2 - V_0^2\right) \qquad (4.3.13)$$

where $V_1^2 = (\nabla \varphi_1)^2$. Representing $\xi = a\sin(kx - \omega t)$ and substituting (4.3.8), and (4.3.9) into the boundary conditions (4.3.11), (4.3.12) and (4.3.13), Kelvin obtained a system of three linear homogeneous algebraic equations with respect to the coefficients A_1, A_2 and a (the terms containing A_1^2 and A_2^2 are neglected). Equating the determinant of this system to zero, he finds the dispersion equation whose solution is

$$\omega = \frac{\rho_2 k V_0}{\rho_1 + \rho_2} \pm \left[\frac{kg(\rho_2 - \rho_1)}{\rho_1 + \rho_2} - \frac{\rho_1 \rho_2 k^2 V_0^2}{(\rho_1 + \rho_2)^2} + \frac{\alpha k^3}{\rho_1 + \rho_2}\right]^{1/2} \qquad (4.3.14)$$

For $\alpha = 0$, $g = 0$, Kelvin's solution (4.3.14) transforms to Helmholtz' solution (4.3.12), and when $V_0 = 0$ and $\alpha = 0$ gives the dispersion equation (4.3.3). From (4.3.14) it follows that the stability condition of capillary-gravity waves on water surface is the non-negativity of the expression in the square brackets, i.e.

$$\frac{\alpha k^2}{\rho_1 + \rho_2} - \frac{\rho_1 \rho_2 V^2}{(\rho_1 + \rho_2)^2} k + \frac{g(\rho_2 - \rho_1)}{\rho_1 + \rho_2} \geq 0 \qquad (4.3.15)$$

By solving inequality (4.3.15) with respect to k, we find that the negativity of its discriminant gives condition of stability of tangential discontinuity for any k in the form

$$V_0^2 \leq \frac{2(\rho_1 + \rho_2)}{\rho_1 \rho_2} \sqrt{\alpha g(\rho_2 - \rho_1)} \qquad (4.3.16)$$

On the other hand, having solved (4.3.15) with respect to V_0, we find the stability condition in the form

$$V_0^2 \leq \frac{[g(\rho_2 - \rho_1) + \alpha k^2](\rho_1 + \rho_2)}{k \rho_1 \rho_2} \qquad (4.3.17)$$

It can be easily demonstrated that minimum value of the right part of inequality (4.3.17) is reached when $k = k_0 = [g(\rho_2 - \rho_1)/\alpha]^{1/2}$ and equals to the right part of inequality (4.3.16), however these conditions contradict to each other. Indeed, from (4.4.16) it follows that if $V_0 \neq 0$ the tangential discontinuity can be stable only in presence of both factors – the surface tension and the gravity field while according to (4.3.17) each of the parameters g and α makes its own contribution to the stability of the tangential discontinuity as it may be stable if one of them is absent. These contradiction which were first detected in the work [17]. quite enough to be convinced that the Kelvin theory is erroneous; however relation (4.3.1) is considered classic up to now.

The another limiting case- long gravity waves (waves on shallow water) is considered in monograph [3] (See §12). The authors solve the problem of wave propagation on water surface in gravity filed of Earth along the canal (along the axis X) with depth h and width b when $V_0 = 0$. Having applied the equations (1.1.1) and (1.1.2) the authors obtain phase speed of wave on the shallow water in the form

$$U_p = \frac{\omega}{k} = \sqrt{gh} \qquad (4.3.18)$$

which coincides with the expression (4.3.4). The equation (1.1.3) is not used at all, assuming that it is satisfied identically. As it is shown in the papers [14] and [22], this is the fundamental error which leading to incorrect solution of the problem on the whole. Indeed, the authors assume that v_z is so small that $\partial v_z / \partial t = 0$. On this assumption, they write down the x and z components of Euler's equation (1.1.1) in the form

$$\frac{\partial v_x}{\partial t} = -\frac{1}{\rho}\frac{\partial p}{\partial x} \qquad (4.3.19)$$

$$\frac{1}{\rho}\frac{\partial p}{\partial z} = -g \qquad (4.3.20)$$

Thereafter they integrated the equation (4.3.20) subject to

$$p\big|_{z=\xi} = p_0 \qquad (4.3.21)$$

and found

$$p = p_0 + \rho g(\xi - z) \qquad (4.3.22)$$

Without going into further details, abovesaid is quite sufficient to get sure that the ultimate result (4.3.18) here is also obtained by means of incorrect solution of the problem:

Firstly – the equation (4.3.20) is nothing else but the fluid equilibrium condition in the Earth gravity field (1.1.12), which is fair only for stationary values of pressure and density and consequently, subject to fulfillment of this condition oscillations are impossible.

Secondly – from condition (4.3.21) it follows that pressure is constant on the disturbed liquid surface and thus, it is unclear how the wave is propagated in this case.

Thirdly- the requirement $\partial v_z / \partial t = 0$ equals to the requirement $v_z = 0$ signifying that liquid surface is not displaced along the axis Z. Notwithstanding this the authors mark this displacement through ξ and obtain wave equation for it.

Solution of the task of shallow water in monograph [3] (see §12) seems to be the more correct. Assuming that $\alpha = 0$ and $V_0 = 0$, and considering motion to be potential, the authors find the dispersive equation for gravity waves on interface of two liquids, density and depth of which equals to ρ_1, h_1 and ρ_2, h_2 respectively, in the form

$$\omega^2 = \frac{kg(\rho_2 - \rho_1)}{\rho_2 cth(kh_2) + \rho_1 cth(kh_1)} \qquad (4.3.23)$$

Three cases are considered:
Both liquids are infinitesimally deep ($kh_1 \gg 1$ and $kh_2 \gg 1$) and then the equation (4.3.23) gives

$$\omega^2 = kg \frac{\rho_2 - \rho_1}{\rho_2 + \rho_1} \qquad (4.3.24)$$

2. Depth of both liquids is small ($kh_1 \ll 1$ and $kh_2 \ll 1$), and then we have

$$\omega^2 = k^2 \frac{g(\rho_2 - \rho_1) h_1 h_2}{\rho_1 h_2 + \rho_2 h_1} \qquad (4.3.25)$$

3. The lower liquid is shallow while the upper liquid is deep ($kh_2 \ll 1$ and $kh_1 \gg 1$) which results in

$$\omega^2 = k^2 g h_2 \frac{\rho_2 - \rho_1}{\rho_2} \qquad (4.3.26)$$

For $\rho_2 \gg \rho_1$, (4.3.24) and (4.3.26) gives the results (4.3.3) and (4.3.4) respectively. Mathematical correctness of solution of this task is conditioned by absence of relative motion of liquids (V_0=0). However, incorrectness related to physics remains in force and consequently it can be claimed that solution (4.3.23) is erroneous.

The basic reason of erroneousness of the Kelvin theory is violation of condition of potentiality of fluid motion, according to which motion can be potential only in isentropic medium. Due importance of this statement, for his evidence, we present calculations of §8 monograph [3].

Circulation of the velocity vector $\vec{V}$ along a closed contour is called the integral

$$\Gamma = \oint_C \vec{V} d\vec{l} \qquad (4.3.27)$$

. Let's consider a closed contour, conducted in a fluid, which consists of its particles. Over time, these particles move, and the entire contour moves with them. Let's denote the differential by coordinates through δ and by time through d. It is clear that $d\vec{l} = \delta\vec{r}$ and then, for changing the circulation we obtain

$$\frac{d}{dt}\oint_C \vec{V}\delta\vec{r} = \oint_C \frac{d\vec{V}}{dt}\delta\vec{r} + \oint_C \vec{V}\frac{d\delta\vec{r}}{dt} \qquad (4.3.28)$$

Since $\vec{V}d(\delta\vec{r})/dt = \vec{V}\delta(d\vec{r}/dt) = \vec{V}\delta\vec{V} = \delta(V^2/2)$ and the integral along a closed contour from the total differential is zero, from (4.4.28), we have

$$\frac{d}{dt}\oint_C \vec{V}\delta\vec{r} = \oint_C \frac{d\vec{V}}{dt}\delta\vec{r} \qquad (4.3.29)$$

Differential of thermal function $w(s, p)$ is

$$dw = Tds + \upsilon dp \qquad (4.3.30)\backslash$$

Let the fluid be isentropic, i.e. $s = const$ at any point. Then for liquid of unit mass $(\upsilon = 1/\rho)$ we obtain

$$dw = \upsilon dp = (1/\rho)dp \qquad (4.3.31)$$

and therefore

$$\frac{1}{\rho}\nabla p = \nabla w \qquad (4.3.32)$$

By using (4.3.32) from equation (1.1.1) at $g = 0$ we obtain the equation of motion for homogeneous liquid in the form

$$\frac{\partial \vec{V}}{\partial t} + (\vec{V}\nabla)\vec{V} = \frac{d\vec{V}}{dt} = -gradw \qquad (4.3.33)$$

Substituting (4.3.33) into (4.3.29) and applying Stokes' formula, we will get

$$\oint_C \frac{d\vec{V}}{dt}\delta\vec{r} = \frac{d}{dt}\oint_C \vec{V}d\vec{l} = \int_S \left(rot\frac{d\vec{V}}{dt}\right)\delta\vec{f} = -\int_S rotgradw\,\delta\vec{f} = 0 \qquad (4.3.34)$$

or

$$\oint_C \vec{V}d\vec{l} = const. \qquad (4.3.35)$$

Thus, we arrive at the result that the circulation of the velocity vector in an ideal homogeneous fluid remains constant in time. This statement is called Thomson's theorem *(W. Thomson, 1869)*, or the law of conservation of velocity circulation. Converting (4.3.35) by Stokes' formula, we obtain

$$\oint_C \vec{V} d\vec{l} = \int_S (ros\vec{V}) d\vec{f} = const \qquad (4.3.36)$$

This implies that if the initial time $rot\vec{V} = 0$, this equality is maintained during infinite time. Such fluid motion is called irrotational, or potential. Since $rot\,grad\varphi \equiv 0$ where φ is any scalar function, we can always define it so that the following equality is fulfilled:

$$\vec{V} = grad\varphi \qquad (4.3.37)$$

Scalar function φ is called the potential of speed $\vec{V}$.

We have emphasized that Thomson's theorem is valid only for the isentropic fluid, i.e. for liquid which is free from the influence of external force field, when the entropy has the same value in all its points. In the gravitational field of the Earth, this condition is violated because of influence of gravitational field, and thus the movement of fluids in this case, by no means, can not be potential. This fact is repeatedly noted in monograph [3] (see §8, §9), however how strange it may seem, in all tasks where the influence of gravitational field is taken into account the condition of potentiality of fluid motion is used. Furthermore in the monograph [19] is proved Helmholtz theorem from which it follows that the circulation of velocity in the gravitational field of the Earth is maintained. We will present this proof fully and will show that it is wrong.

Circulation of speed $\vec{V}$ the author writes in the form

$$\Gamma(t) = \oint_C V_x dx + V_y dy + V_z dz = \oint_C V_l dl \qquad (4.3.38)$$

where V_l is the projection of the velocity vector on the tangents which is held at some point of closed liquid contour, and dl is a length of element of contour containing this point. The contour is defined parametrically by vector $\vec{X}(\sigma,t)$, where $0 \le \sigma \le 1$, and $\vec{X}(0,t) = \vec{X}(1,t)$. Thus, the selection of parameter σ means selection of certain liquid particle on the contour. In this case, formula (4.3.38) can be written as

$$\Gamma(t) = \int_0^1 \vec{V} \frac{\partial \vec{X}}{\partial \sigma} d\sigma \qquad (4.3.39)$$

Derivative (4.3.39) by time is

$$\frac{d\Gamma}{dt} = \int_0^1 \left(\frac{d\vec{V}}{dt} \frac{\partial \vec{X}}{\partial \sigma} + \vec{V} \frac{d}{dt} \frac{\partial \vec{X}}{\partial \sigma} \right) d\sigma = \int_0^1 \left(\vec{a} \frac{\partial \vec{X}}{\partial \sigma} + \vec{V} \vec{V}_\sigma \right) d\sigma \qquad (4.3.40)$$

where $\vec{a} = d\vec{V}/dt$ is acceleration of particles of liquid and $\vec{V}_\sigma = d(\partial \vec{X}/\partial \sigma)/dt = \partial \vec{V}/\partial \sigma$ is component of the velocity of the particle along the tangent of the contour. Let's write Euler's equation of motion of the fluid in the form of Lagrange

$$\vec{a} = \frac{d\vec{V}}{dt} = -\frac{\nabla p}{\rho} + \vec{g} = -\frac{\nabla p}{\rho} - g\nabla z \qquad (4.3.41)$$

Substitution of (4.3.41) into (4.3.40) gives

$$\frac{d\Gamma}{dt} = \int_0^1 \left[-\frac{1}{\rho}\frac{\partial \vec{X}}{\partial \sigma}\nabla p - g\frac{\partial \vec{X}}{\partial \sigma}\nabla z + \vec{V}\frac{\partial \vec{V}}{\partial \sigma} \right] d\sigma \qquad (4.3.42)$$

However, $(\partial \vec{X}/\partial \sigma)\nabla P = \partial P/\partial \sigma$, $(\partial \vec{X}/\partial \sigma)\nabla z = \partial z/\partial \sigma$, $\vec{V}(\partial \vec{V}/\partial \sigma) = (1/2)(\partial V^2/\partial \sigma)$ and from (4.3.42), we obtain

$$\frac{d\Gamma}{dt} = \int_0^1 \left[-\frac{1}{\rho}\frac{\partial p}{\partial \sigma} - g\frac{\partial z}{\partial \sigma} + \frac{1}{2}\frac{\partial V^2}{\partial \sigma} \right] d\sigma = -\frac{1}{\rho}[p(0) - p(1)] + g[z(0) - z(1)] + \frac{1}{2}[V^2(1) - V^2(0)] \quad (4.3.43)$$

According to the author, the expression (4.3.43) is equal to zero, because in consequence of cyclical of parameter σ, the following equations: $P(0) = P(1)$, $z(0) = z(1)$ and $V(0) = V(1)$ are met. This is a profound error, since the cyclical parameter determines the point on the contour but not physical quantities at this point. These equalities are executed only if the points of contour (liquid particles) move in a strictly horizontal direction at a constant speed. Thus, the circulation of velocity in a closed contour in inhomogeneous (nonisentropic) fluid preserved under steady motion. Any deviation of the liquid particles along the Z axis causes a change the speed circulation and movement is nonpotential. Indeed, in the gravity field of Earth the z component of disturbed velocity must be depend on gravity acceleration g, which must be taken into account in potentials of velocities (4.3.8) and (4.3.9). However, in this case implementation of condition of potentiality of fluid motion $\partial v_x/\partial z = \partial v_z/\partial x$ is impossible to be met and consequently, in such setting the problem can not be solved. That is why the constant term $-\rho_1 V_0^2/2$ is artificially introduced in the right side of the equation (4.3.13).

It should be paid attention to one circumstance. At horizontal motion ($z(0) = z(1)$), when circulation of velocity is preserved ($d\Gamma/dt = 0$), relation (4.3.43) gives

$$\frac{1}{2}\nabla V^2 = (\vec{V}\nabla)\vec{V} = -\frac{1}{\rho}\nabla p \qquad (4.3.44)$$

which coincides with the stationary motion equation ($\partial \vec{V}/\partial t = 0$), i.e. such movement really is potential. On the other hand, as shown above, the speed of sound in water is isobaric i.e water is always compressible and therefore $\nabla \vec{V} \neq 0$, which means that the potential of velocity does not satisfy Laplace equation $\Delta \varphi = 0$. Indeed, as follows from (1.1.7) and (1.1.9)

$$\frac{d\rho}{dt} = \frac{1}{C_s^2}\frac{dp}{dt} \qquad (4.3.45)$$

and the equation of mass continuity (2.2.6) gives

$$\frac{d\rho}{dt} = -\rho \nabla \vec{V} \qquad (4.3.46)$$

Comparing these two equations we easily find that

$$\nabla \vec{V} = -\frac{1}{\rho C_s^2}\frac{dp}{dt} \qquad (4.3.47)$$

from which, for potential of velocity we get Poisson equation

$$\Delta \varphi = -\frac{1}{\rho C_s^2}\frac{dp}{dt} \qquad (4.3.48)$$

Thus, the velocity potential can satisfy the equation of Laplass if $dP/dt = 0$ or when $|dp/dt| \ll |\rho C_s^2|$.

§ 4. The correct solution of the task of the capillary waves.

Let us now consider the correct linear theory of surface waves with consideration of forces of surface tension and gravity. Above we showed that in the atmosphere at sea level as well as in water $C_p = \infty$ and $C = C_s$. Let the air move relatively to the water, which has a flat bottom $z = -h$ with speed $\vec{V}_0 = V_0 \vec{e}_x$ and let's assume that $\nabla \rho_0(z) = \partial \rho_0/\partial z$ is a small quantity. Then for perturbed pressure in the air, we can write the equation (3.1.12) (see formula (3.3.3)) in the form

$$\frac{1}{\widetilde{C}_1^2}\frac{\partial^2 p_1}{\partial t^2} - \frac{g}{\widetilde{C}_1^2}\frac{\partial p_1}{\partial z} - \Delta p_1 = V_0\left[\rho_{01}\frac{\partial \nabla \vec{V}_1}{\partial x} - \frac{1}{\widetilde{C}_1^2}\frac{\partial^2 p_1}{\partial x \partial t}\right] \qquad (4.4.1)$$

Expressing $\nabla \vec{V}_1$ from the linearized equation of continuity of mass (4.1.14) and substituting in (4.4.1), we obtain

$$\frac{1}{\widetilde{C}_1^2}\frac{\partial^2 p_1}{\partial t^2} - \frac{g}{\widetilde{C}_1^2}\frac{\partial p_1}{\partial z} - \Delta p_1 = -\frac{V_0}{\widetilde{C}_1^2}\left[\frac{\partial}{\partial x}\left(2\frac{\partial p_1}{\partial t} + V_0\frac{\partial p_1}{\partial x}\right)\right] \qquad (4.4.2)$$

Equation (4.4.2) coincides with Landau equation (4.1.15) when $g = 0$. We seek a solution of equation (4.4.2) in the form (4.1.3), then for the amplitude of the perturbed pressure obtain the equation

$$\frac{d^2 p_{a1}}{dz^2} + \frac{g}{\tilde{C}_1^2}\frac{dp_{a1}}{dz} + \frac{1}{\tilde{C}_1^2}\left[k^2\tilde{C}_1^2 - (\omega - kV_0)^2\right]p_{a1} = 0 \qquad (4.4.3)$$

Solution of the equation (4.4.3) with consideration of attenuation of waves when $z \to \infty$ (surface wave) is

$$p_{a1}(z) = A\exp(\gamma k z)$$

(4.4.4)

where

$$\gamma = -\frac{1}{\theta_1}\left[1 + \sqrt{1 + \theta_1^2\left(1 - \frac{(U_p - V_0)^2}{\tilde{C}_1^2}\right)}\right] < 0 \qquad (4.4.5)$$

$$\theta_1 = \frac{2k\tilde{C}_1^2}{g} \qquad (4.4.6)$$

and $U_p = \omega/k$.

For water the equation (3.3.3) will be as follows:

$$\Delta p_2 + \frac{g}{C_2^2}\nabla p_2 - \frac{1}{C_2^2}\frac{\partial^2 p_2}{\partial t^2} = 0 \qquad (4.4.7)$$

Solution of equation (4.5.7) for amplitude of pressure disturbance in water gives

$$p_{a2}(z) = B_1\exp(\delta_1 k z) + B_2\exp(\delta_2 k z) \qquad (4.4.8)$$

where

$$\delta_1 = -\frac{1}{\theta_2}\left[1 - \sqrt{1 + \theta_2^2\left(1 - \frac{U_p^2}{C_2^2}\right)}\right] > 0 \quad, \qquad (4.4.9)$$

$$\delta_2 = -\frac{1}{\theta_2}\left[1 + \sqrt{1 + \theta_2^2\left(1 - \frac{U_p^2}{C_2^2}\right)}\right] < 0 \qquad (4.4.10)$$

$$\theta_2 = \frac{2kC_2^2}{g} \qquad (4.4.11)$$

Thus, for disturbed values of pressure in air and water we will have

$$p_1(x,z,t) = A\exp(\gamma kz)\exp[i(kx-\omega t)] \qquad (4.5.12)$$

$$p_2(x,z,t) = [\,B_1\exp(\delta_1 kz) + B_2\exp(\delta_2 kz)\,]\exp[i(kx-\omega t)] \qquad (4.5.13)$$

After use of equilibrium condition (3.1.4) and medium state equation (3.1.5) when $C = C_s$ the linearized Euler's equation (the first equation of system (3.1.3)) will be as follows:

$$\frac{\partial \vec{V}}{\partial t} + (\vec{V}_0 \nabla)\vec{V} = -\frac{1}{\rho_0}\nabla p + \frac{p}{\rho_0 \tilde{C}^2}\vec{g} \qquad (4.4.14)$$

Having represented in (4.4.14) all variables in the form (4.1.3), for amplitude of the z component of disturbance velocity we will get

$$V_{az}(z) = -\frac{i}{k\rho_0(U_p - V_0)}\left[\frac{dp_a(z)}{dz} + \frac{p_a(z)}{\tilde{C}^2}g\right] \qquad (4.4.15)$$

From (4.4.15) for air and water respectively we will have

$$V_{1z}(x,z,t) = -\frac{i}{k\rho_{01}(U_p - V_0)}\left[\frac{dp_{a1}(z)}{dz} + \frac{p_{a1}(z)}{C_1^2}g\right]\exp[i(kx-\omega t)] \qquad (4.4.16)$$

$$V_{2z}(x,z,t) = -\frac{i}{k\rho_{02}U_p}\left[\frac{dp_{a2}(z)}{dz} + \frac{p_{a2}(z)}{C_2^2}g\right]\exp[i(kx-\omega t)] \qquad (4.4.17)$$

Here, we believe that $C_1 \gg V_0$ and therefore $\tilde{C}_1 = C_1$. As we see disturbance velocity in area $z > 0$ as well as in area $z < 0$ obviously depends on g, which does not occur in Kelvin theory (see formulas (4.3.8) and (4.3.9)). Now, to boundary conditions (4.3.10), (4.3.11) and (4.3.12), should be added the condition on the bottom of water reservoir

$$V_{2z}\big|_{z=-h} = 0 \qquad (4.4.18)$$

and substituting in this conditions the values of relevant quantities from formulae , (4.1.6), (4.4.12), (4.4.13), (4.4.16) , (4.4.17) , we will get the system of linear homogeneous algebraic equations regarding coefficients A, B_1. B_2 and ξ_a. Equating the determinant of this equation to zero, we shall find dispersive equation of linear surface capillary-gravity waves in the form of

$$[\rho_{01}(U_p - V_0)^2 - \gamma^* \alpha k][\exp(-\delta_2 kh) - \exp(-\delta_1 kh)]\delta_1\delta_2 - \rho_{02}U_p^2\gamma^*[\delta_1 \exp(-\delta_2 kh) - \delta_2 \exp(-\delta_1 kh)] = 0 \qquad (4.4.19)$$

where

$$\gamma^* = -\frac{1}{\theta_1}\left[1 - \sqrt{1 + \theta_1^2\left(1 - \frac{(U_p - V_0)^2}{C_1^2}\right)}\right] > 0 \qquad (4.4.20)$$

Let us consider condition $\theta_1 > 1$. Taking into consideration that at sea level $C_1 \approx 340$ m/sec and $g \approx 10$ m/sec^2 we shall find $k > 8{,}64 \times 10^{-5}$ m^{-1} or $\lambda < 0{,}72 \times 10^5$ m. Thus, this condition covers the whole range of wavelengths from capillary to tsunami and the bigger is k the better it is met. For instance, when the value of gravity wave length is $\lambda \approx 100$ m, for air $\theta_1 \approx 0{,}71 \times 10^3 \gg 1$ while for water $\theta_2 \approx 0{,}14 \times 10^5 \gg 1$. It is also apparent that $U_p^2/C_{s2}^2 \ll U_p^2/C_{s1}^2 \ll 1$ and then it is easy to check that $\delta_1 = \gamma^* = 1$ and $\delta_2 = -1$, following which dispersive equation (4.4.19) is simplified and takes the form

$$[\rho_{01}(U_p - V_0)^2 - \alpha k]th(kh) = -\rho_{02}U_p^2 \qquad (4.4.21)$$

As we see, the equation (4.4.21) does not contain gravity acceleration g and does not have solution when $\alpha = 0$, thus it can be concluded that linear theory of surface capillary-gravity waves is adequate only for capillary waves on which gravity field has no effect. Furthermore, it is obvious that when $\alpha = 0$ and $g = 0$, the task loses its physical meaning and therefore the solution of Helmholtz (4.1.12) does not contain any information. We also see that a necessary (but not sufficient) condition for the existence of real roots of the equation (4.4.21) is fulfilment of inequality

$$(U_p - V_0)^2 < \frac{\alpha k}{\rho_{01}} \Rightarrow \omega < k\left(V_0 \pm \sqrt{\frac{\alpha k}{\rho_{01}}}\right) \qquad (4.4.22)$$

whereupon these roots are equal

$$\omega = k\frac{\rho_{01}V_0 th(kh) \pm \sqrt{th(kh)\{[\rho_{01}th(kh) + \rho_{02}]\alpha k - \rho_{01}\rho_{02}V_0^2\}}}{\rho_{01}th(kh) + \rho_{02}} \qquad (4.4.23)$$

In our case, the inequality $\rho_{01}th(kh) \ll \rho_{02}$ is right and then from (4.4.23), we obtain

$$\omega = k\frac{\rho_{01}V_0 th(kh) \pm \sqrt{th(kh)[\rho_{02}\alpha k - \rho_{01}\rho_{02}V_0^2]}}{\rho_{02}} \qquad (4.4.24)$$

and therefore, the stability condition of capillary waves will be

$$V_0^2 \le \frac{\alpha k}{\rho_{01}} \qquad (4.4.25)$$

Condition (4.4.25) coincides with condition (4.3.17) at $g=0$ and $\rho_2 \gg \rho_1$ that proves once again that the movement could be potential only in a homogeneous (isentropic) environment. From this condition we can easily calculate wind speed at which capillary wave of a certain length would be blown away. For example, for the water $\alpha = 0{,}073$ N / m while for the air $\rho_{01} = 1{,}2$ kg / m^3, and then from (4.5.25), we obtain

$$\lambda \le \frac{0{,}4}{V_0^2} \qquad (4.4.26)$$

Thus, the wind speed of which is equal to $V_0 = 5$ m / sec will blow away capillary waves, length of which $\lambda > 1{,}6$ cm.

At $V_0 = 0$ the solution (4.4.24) gives

$$U_p = \frac{\omega}{k} = \pm\sqrt{\frac{k\alpha}{\rho_{02}}\,th(kh)} \qquad (4.4.27)$$

This solution invalidates the present opinion that capillary waves are generated only in deep water. We see that they are generated in deep ($kh > 1, th(kh) \approx 1$) as well as in shallow ($kh < 1, th(kh) \approx kh$) water. In the first instance dispersive equation is given by

$$\omega = \pm k\sqrt{\frac{k\alpha}{\rho_{02}}} \qquad (4.4..28)$$

and in the second one

$$\omega = \pm k^2\sqrt{\frac{\alpha h}{\rho_{02}}} \qquad (4.4.29)$$

Since condition (4.3.2) who limiting the length of capillary waves no longer exists, let us consider perturbation with wavelength $\lambda = 0{,}1$ m ($k = 62{,}8$ m^{-1}) for which from formulae (4.4.28) and (4.4.29) we will obtain that in deep water ($h \ge 0{,}5$ m) $\omega = 0{,}13$ sec^{-1} and $U_p = 2$ cm/sec and in shallow water ($h = 0{,}05$ m) $\omega = 0{,}07$ sec^{-1} and $U_p = 1$ cm/sec. As we see, frequency and phase speed of capillary wave drop two times when depth lowers 10 times.

The results of this section were published in the work [27]

Results:

Based on the correct formulation and solution of the linear problem surface waves in the Earth's gravitational field, in this chapter the following new results are obtained:

- In the gravitational field of the Earth only stationary motion of the liquid may be potential on the condition that all the fluid particles move in a strictly horizontal direction.
- It is shown that the formulation and solution of task of linear surface waves in fluids does not make sense without taking into account of force of surface tension.
- Solutions of Helmholtz (1868) and Landau (1944) and the following from them the conditions of the stability of a tangential discontinuity in incompressible and compressible fluids which are considered as classic are contradictory and do not bear any information.
- In solving by Kelvin the linear problem of capillary-gravitational waves (1871) was made scientific incorrectness which initiated the existing contraditictions
- The linear theory of surface waves describes only the capillary waves, influence on which the Earth's gravitational field is negligibly small and, therefore, capillary-gravity waves do not exist. Consequently, there is no condition which limits the length of the capillary waves.
- The capillary waves can be generated and spread not only in deep, but also in shallow water, where the phase velocity of the wave depends on the depth of water bodies.
- The potential of velocity of the fluid satisfies not Laplace equation, but Poisson equation.
- The formula is obtained which can be used to calculate the maximum length of capillary waves at a given speed of wind.

Conclusion

This book contains the author's all works, published by him for the last four years. These publications were very difficult, for they denied the beliefs, entrenched over the last two centuries, on fundamental values and criteria of gas dynamics, hydrodynamics, and even thermodynamics. So for example, the sound speed, which is the most important characteristic of medium, plays the main part in all problems of gas and hydrodynamics, was determined without possible heterogeneous medium, and scientists still use this definition both in homogeneous and heterogeneous media. The reason is that the air density distribution in the Earth's atmosphere is such that on the sea-level, gravitational acceleration does not affect the generation and propagation of sound waves, and so, at low altitudes ($z < 2,3 km$) in relation to the sound waves in the atmosphere can be considered as a homogeneous medium, free from the influence of gravitational field of the Earth. On this basis, the researchers of the Earth's atmosphere do not doubt the validity of the formula determining the speed of sound, following which the sound speed depends only on temperature, but does not depend on the atmospheric density. Our research strongly prove that along with adiabatic mechanism of generation and propagation of sound in the atmosphere of Earth, which in modern physics is considered to be the only mechanism, there is one more mechanism – isobaric. Each of them has its own propagation speed, which we call adiabatic (C_s) and isobaric (C_p) sound speeds. The true value of the square of the sound speed is reduced from the squares of these two speeds, and so as isobaric speed obviously depends on the altitude, the sound speed also depends on altitude or what is the same thing, from density. The greater is the altitude, the stronger is this dependence. For example, at a height of 1 km, the relative error of calculations of the existing and obtained by us formulas is only 0,3%, whereas at the height of 11 km., which is the average height of the upper boundary of the troposphere, it is 100 times more. This height can be determined with good accuracy from the equation $C_s = C_p$ and it means that on this altitude frequencies of adiabatic and isobaric sound vibrations are equalized, i.e. a resonance takes place. This conclusion is confirmed by experimental graph (Fig.1) which shows that the monotonous fall of temperature with the growing altitude from sea level, at this altitude it is interrupted abruptly and then it remains constant until the upper boundary of the tropopause (20km) and then increases approximately linearly up to the upper boundary of the stratosphere (50 km). Such dynamic of the temperature changes shows that the resonance phenomenon contributes to a sharp increase the intensity of the absorption of solar radiation (presumable infrared radiation), and therefore our theory is valid only up to the upper boundary of the troposphere, because above this the Earth's atmosphere, is not adiabatic.

The existence of two mechanisms of sound generation is associated with two ways of weak perturbations of the density of matter: adiabatic change in mass of a substance in a given volume and isobaric change of volume of a given mass. We have shown that in homogeneous media the first method prevails and therefore, the sound is adiabatic ($C_p = \infty$ и $C = C_s$). On the other hand, adiabatic change in mass in a given volume is described by the existing equation of continuity of mass $d\rho/dt = -\rho div \vec{V}$, and therefore, the generally accepted criterion of incompressibility $C_s = \infty \Rightarrow div\vec{V} = 0$ of air at low altitudes is not applicable. For highly heterogeneous medium $C_s = \infty \Rightarrow div\vec{V} = 0$ and isobaric density perturbation is described by the equation $d\rho/dt = -\vec{V}\nabla p/C_p^2$. Thus, only highly heterogeneous medium can be incompressible and only homogeneous medium can be compressible. Based on the foregoing, the terms "compressible medium" and "incompressible medium" which had a purely mechanical sense, acquire the thermodynamic meaning. So for example, we have shown that the water and the iron in the thermodynamic sense are much more compressible than the air at altitudes of $z > 2,3$ km, since the speed of sound in these media is adiabatic and consequently, they are homogeneous media in relation to the sound waves. This result is quite understandable, because the sound in condensed media propagates due to fluctuations of the forces of elasticity acting between the molecules and crystal nodes, exceeding by many times the gravitational force which does not affect these fluctuations.

Another important result of our research is that we expressed isobaric speed of sound in the medium in any aggregate state through its coefficient of thermal expansion. In consequence of this, we found that the value of this coefficient for an ideal gas varies with the changes of altitude, and in different ways for different gases. Thus, we have shown that the opinion of the universality of the laws of ideal gas that has existed in science for over more then two centuries, is incorrect.

Obviously, our arguments and findings can contribute a lot to the modern theory of dynamical processes in liquids and gases. This particularly applies to the theory of hydrodynamic tangential discontinuity, by which solved both linear and nonlinear problems of surface gravity waves. We showed the drawbacks of the linear theory of surface gravity waves, but nonlinear theory is the most important which we do not consider here. The major environmental problems of interaction of the ocean and the coast, is the problems of waves of tsunami, who can be solved using this theory. A huge number of scientific papers (see, for example [3],[9],[20]), are dedicated to this issue, and they are based on the assumption of incompressibility of water and the potentiality of its movement. We have shown that these assumptions are not applicable to water and therefore, the existing nonlinear theory of surface gravitational waves demands revision.

The incorrectness of the decision of problems of hydrodynamic tangential discontinuity for ordinary fluids has left its mark on the solution of similar problems in magnetic hydrodynamics. Conditions for the stability of the magnetohydrodynamic (MHD) tangential discontinuity for compressible and incompressible conductive liquids also contradict each other as in ordinary liquids. For example, according to the calculations by Syrovatski [25], the relative period of two incompressible conducting liquid along a surface of tangential discontinuity leads to its destabilization, while magnetic field intensities on both sides of this surface stabilize it. According to calculations by Sen [28] in a compressible conductive liquids, it is opposite: the greater is the relative velocity of flow, and the less is the intensity of the magnetic field, the MHD tangential discontinuity is more stable. These issues are discussed in the work [24] in detail.

In conclusion, we can say that this book will enable professionals rethink the applied problems of hydrodynamics and magnetohydrodynamics.

REFERENCES

[1] U.S. Standard Atmosphere, National Aeronautics and Space Administration, 1976.
[2] http://www.digitaldutch.com/atmoscalc
[3] L. D. Landau and E. N. Lifshitz, "Nauka," Theoretikal Physics, Hydrodynamics, Vol. 6, Moscow, 1988.
[4] A. D. Pirce, "Acoustics: AnIntroduction to Its Physical Principles and Applications," Acoustical Society of America, New York, 1989.
[5] E. E. Gossard and W. H. Hooke, "Waves in the Atmosphere," Elsevier, New York, 1975.
[6] L. M. B. C. Campos, "On Three-Dimensional Acoustic Gravity Waves in Model Non-Isothermal Atmospheres," Wave Motion, Vol. 5, No. 1, 1983, pp. 1-14. doi:10.1016/0165-2125(83)90002-1
[7] M. J. Lighthill, "Waves in Fluids," Cambridge University Press, Cambridge, 2002.
[8] G. S. K. Wang, "Speed of Sound in Standard Air," Journal of the Acoustical Society of America, Vol. 79, No. 5, 1986, pp. 1359-1366. doi:10.1121/1.393664
[9] G. Santostasi, et al., "A Student Designed Experiment Measuring the Speed of Sound as a Function of Altitude," McNeese State University, Lake Charles, 2008.
[10] V.G. Kirtskhalia, „Speed of Sound in Atmosphere of the Earth", Open Journal of Acoustics, (2012), 2, 80-85, doi:10.4236/oja.2012.22009
[11] http://www.engineeringtoolbox.com/air-properties-d_156.html
[12] L.D. Landau and E.M. Lifshitz , Statistical Physics, Publishing House "MIR" Moscow, 1964.
[13] V.G. Levich, Course of Theoretical Physics, Vol. I, Publishing House "FIZMATGIS", Moscow, 1962.
[14] V.G. Kirtskhalia, "The New Determination of the Criteria of Compressibility and Incompressibility of Medium", Journal of Modern Physics, (2013) , V.4, No. 8, p.p. 1075-1079, doi: org/10.4236/jmp.2013.48144
[15] V.G. Kirtskhalia, " On universality of laws of ideal gas", Journal of Modern Physics, V.6, No. 7, p.p. 448-454, doi: 10.4236/jmp.2015.67099 .
[16] V. G. Kirtskhalia and A. A. Rukhadze, "The Influence of Effective Gravity Field on the Development of Instability Tangential Discontinuity," Kratkie Soobshchenya po Fizike, No. 4, Moscow, 2006.
[17] V. G. Kirtskhalia and A. A. Rukhadze, "On the Question of Hydrodynamic Tangential Gap," Georgian International Journal of Science and Technology, Vol. 1, No. 3, 2008.
[18] A. F. Aleksandrov, L. S. Bogdankevich and A. A. Rukhadze, "Osnovi Electrodinamiki Plazmi," Moscow, 1978.
[19] J.J. Stoker, "Water Waves", Interscienjse publishers, Inc., New York, 1957
[20] Z. Kowalik, "Introduction to Numerical Modeling of Tsunami Waves", Institute of Marine Science University of Alaska, Fairbanks, 2012.
[21] http://en.wikipedia.org/wiki/Doppler_effect
[22] V. G. Kirtskhalia, "Sound wave as a particular case of the gravitational wave", Open Journal of Acoustics, (2012), V. 2, No. 3, p.p. 115-120. doi.org/10.4236/oja.2012.23013
[23] V. G. Kirtskhalia , "Peculiarities of Sound Wave in Moving Medium", Open Journal of Acoustics, (2014), V.4, 99-104. http://dx.doi.org/10.4236/oja.2014.43010
[24] V. G. Kirtskhalia, "On the stability problem of the tangential discontinuity" Planet. Space Sci. Vol. 42, No.6, pp.513- 518, 1994. doi:10.1016/0032-0633(94)00083-2
[25] Syrovatsky, S. I., Zh. Eksperiment. i Teor. Fiziki 24, 622, 1953.

[26] R. A. Gerwin, " On the stability problem of the tangential discomtinuity"
Rev. Mod. Phys. 40, 652,1968.
[27] V. G. Kirtskhalia, „On Gravity Waves on the Surface of Tangential Discontinuity" ,
 Applied Physics Research; Vol. 6, No. 2; p.p. 109-117, 2014, doi:10.5539/apr.v6n2p109
[28] A. K. Sen, " Effect of compressibility on Kelvin-Helmholtz instability in a plasma", Phys. Fluids 7, p. 1293, 1964.

I want morebooks!

Buy your books fast and straightforward online - at one of the world's fastest growing online book stores! Environmentally sound due to Print-on-Demand technologies.

Buy your books online at
www.get-morebooks.com

Kaufen Sie Ihre Bücher schnell und unkompliziert online – auf einer der am schnellsten wachsenden Buchhandelsplattformen weltweit!
Dank Print-On-Demand umwelt- und ressourcenschonend produziert.

Bücher schneller online kaufen
www.morebooks.de

OmniScriptum Marketing DEU GmbH
Heinrich-Böcking-Str. 6-8
D - 66121 Saarbrücken
Telefax: +49 681 93 81 567-9

info@omniscriptum.com
www.omniscriptum.com

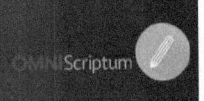

www.ingramcontent.com/pod-product-compliance
Lightning Source LLC
Chambersburg PA
CBHW031543210526
45464CB00003B/1129